Lecture Notes in Mechanical Engineering

Lecture Notes in Mechanical Engineering (LNME) publishes the latest developments in Mechanical Engineering - quickly, informally and with high quality. Original research reported in proceedings and post-proceedings represents the core of LNME. Volumes published in LNME embrace all aspects, subfields and new challenges of mechanical engineering. Topics in the series include:

- Engineering Design
- Machinery and Machine Elements
- Mechanical Structures and Stress Analysis
- Automotive Engineering
- Engine Technology
- Aerospace Technology and Astronautics
- Nanotechnology and Microengineering
- Control, Robotics, Mechatronics
- MEMS
- Theoretical and Applied Mechanics
- Dynamical Systems, Control
- Fluid Mechanics
- Engineering Thermodynamics, Heat and Mass Transfer
- Manufacturing
- Precision Engineering, Instrumentation, Measurement
- Materials Engineering
- Tribology and Surface Technology

To submit a proposal or request further information, please contact the Springer Editor in your country:

China: Li Shen at li.shen@springer.com
India: Dr. Akash Chakraborty at akash.chakraborty@springernature.com
Rest of Asia, Australia, New Zealand: Swati Meherishi at swati.meherishi@springer.com
All other countries: Dr. Leontina Di Cecco at Leontina.dicecco@springer.com

To submit a proposal for a monograph, please check our Springer Tracts in Mechanical Engineering at http://www.springer.com/series/11693 or contact Leontina.dicecco@springer.com

Indexed by SCOPUS. The books of the series are submitted for indexing to Web of Science.

More information about this series at http://www.springer.com/series/11236

Fakher Chaari · Maher Barkallah ·
Anas Bouguecha · Bassem Zouari ·
Mohamed Taoufik Khabou ·
Mounir Kchaou · Mohamed Haddar
Editors

Advances in Materials, Mechanics and Manufacturing

Proceedings of the Second International Conference on Advanced Materials, Mechanics and Manufacturing (A3M'2018), December 17–19, 2018 Hammamet, Tunisia

Springer

Editors
Fakher Chaari
Department of Mechanical Engineering
National School of Engineers of Sfax
Sfax, Tunisia

Maher Barkallah
Department of Mechanical Engineering
National School of Engineers of Sfax
Sfax, Tunisia

Anas Bouguecha
National School of Engineers of Gafsa
Gafsa, Tunisia

Bassem Zouari
Department of Mechanical Engineering
National School of Engineers of Sfax
Sfax, Tunisia

Mohamed Taoufik Khabou
Department of Mechanical Engineering
National School of Engineers of Sfax
Sfax, Tunisia

Mounir Kchaou
Sfax Preparatory Engineering Institute
Sfax, Tunisia

Mohamed Haddar
Department of Mechanical Engineering
National School of Engineers of Sfax
Sfax, Tunisia

ISSN 2195-4356 ISSN 2195-4364 (electronic)
Lecture Notes in Mechanical Engineering
ISBN 978-3-030-24246-6 ISBN 978-3-030-24247-3 (eBook)
https://doi.org/10.1007/978-3-030-24247-3

This Springer imprint is published by the registered company Springer Nature Switzerland AG
The registered company address is: Gewerbestrasse 11, 6330 Cham, Switzerland

Preface

Two years have passed after organizing the first edition of the International Conference on Advanced Materials Mechanics and Manufacturing. The second edition of **A3M** Conference is organized by the Laboratory of Mechanics, Modelling and Manufacturing (LA2MP), National School of Engineers of Sfax, University of Sfax, from 17 to 19 December 2018, at Hammamet, Tunisia. In the two years between the first and second conferences, many research topics in mechanical engineering have seen a considerable evolution both in theory and in practice. Actually, the goal of mechanical designers becomes more easily attainable due to accurate models and efficient decision-support tools in mechanics of materials, smart and complex system physics, and manufacturing and non-destructive testing methods. The aim of this international conference is indeed to bring together academic and industrial researchers to shear their knowledge, to build new relationships and to offer new opportunities susceptible to create research projects dealing with material behaviour, characterization and simulation taking into account their application in manufacturing.

The conference covers several subjects as following:

- Material behaviour: modelling and characterization
- Simulation
- Technologies
- Materials

We would like to express our gratitude to the researchers and scientists who take part in this international event as scientific committee members, plenary session speakers, and oral and poster presenters, as well as to everybody who has contributed to the success of this edition of the conference.

Hammamet, Tunisia
December 2018

Fakher Chaari
Maher Barkallah
Anas Bouguecha
Bassem Zouari
Mohamed Taoufik Khabou
Mounir Kchaou
Mohamed Haddar

Contents

Dynamical Viscoelastic Properties of Poly (Ester-Urethane) Biomaterial for Scaffold Applications

Géraldine Rohman[1,2(✉)], Salah Ramtani[1,2], Sylvie Changotade[1], Credson Langueh[1], Yves Roussigné[3], Florent Tétard[3], Fréderic Caupin[4], and Philippe Djemia[2,3]

[1] Laboratoire Chimie, Structures, Propriétés de Biomatériaux et d'Agents Thérapeutiques CSPBAT UMR7244 CNRS & Université Paris 13, Villetaneuse, France
{geraldine.rohman, ramtani, changotade, credson.langueh}@univ-paris13.fr
[2] Institut Interdisciplinaire des Sciences Expérimentales, Université Paris 13, Villetaneuse, France
{geraldine.rohman, ramtani, djemia}@univ-paris13.fr
[3] Laboratoire des Sciences des Procédés et des Matériaux LSPM UPR 3407 CNRS, Sorbonne Paris Cité, Villetaneuse, France
{yves.roussigne, florent.tetard, djemia}@univ-paris13.fr
[4] Institut Lumière Matière ILM, UMR 5306, Université Lyon, Lyon, France
frederic.Caupin@univ-lyon1.fr

Abstract. Biodegradable poly(ester-urethane)-based scaffolds with an elastomeric character offer special mechanical properties by reducing the mismatch of Young's modulus between rigid thermoplastic scaffolds classically used, and soft and dynamic tissues to regenerate (skin, tendons, muscles). In this study, porous scaffolds had been elaborated by using a high internal phase emulsion process with variable pores size (150–1800 μm), a porosity of 85% and sufficient strength making them suitable for cell culture, tissue formation and therefore various tissue engineering applications. The scaffold mechanical properties were evaluated at different frequencies. Analysis by conventional quasi-static mechanical tests (0.1 Hz) demonstrated that the scaffolds exhibited an elastomeric character with an effective Young modulus of 165 kPa. In addition, the determination of dynamical viscoelastic parameters was assessed at higher frequencies (MHz–GHz) by the ultrasonic pulse echo method (10 MHz) and micro-Brillouin inelastic light scattering (13 GHz) analysis. Our measurements revealed a strong increased of the Young modulus with increasing frequency and a linear correlation (log-log scale) between stiffer hypersonic (MHz–GHz) modulus and softer low-frequency one. The slope $\alpha = 0.194$ is characterizing the viscoelastic behavior of this polymer. In addition, micro-Raman was coupled to the micro-Brillouin to determine at the same location the chemical properties of the polymer.

Keywords: Biomaterial · Poly(ester-urethane) · Scaffold · Viscoelastic · Young modulus

F. Chaari et al. (eds.), *Advances in Materials, Mechanics and Manufacturing*, Lecture Notes in Mechanical Engineering,
https://doi.org/10.1007/978-3-030-24247-3_1

1 Introduction

The purpose of tissue engineering is to develop new materials adapted to the reconstruction site and is based on the combination of cells inoculated into a porous biodegradable structure (scaffolding). The evaluation of mechanical properties in relation to the porosity of scaffolding is paramount since it has been shown that the cells are sensitive to the rigidity and spatial structuring of their microenvironment, and thus to the mechanical properties and the porosity of the scaffold. Therefore, they regulate their form, proliferation, and cortical stiffness according to their adhesion support [10]. Recently, poly(ester-urethane)-based biomaterials are increasingly being used as they can provide scaffolds with an elastomeric character [4]. These elastomer-type biomaterials offer special mechanical properties by reducing the mismatch of the Young modulus between conventional rigid thermoplastic scaffolds and the soft and dynamic tissues to be regenerated (skin, tendons, muscles), and on their ability to recover large mechanical deformations. CSPBAT has developed new biodegradable scaffolds based on poly(ester-urethane) (PCLU) showing their ability to support the adhesion of mesenchymal stem cells [3]. It is necessary to characterize the intrinsic elastic properties of these new elastomer scaffolds and understand the correlation between the porosity of the material and its macroscopic rigidity. In addition to conventional quasi-static mechanical tests, the pulse-echo ultrasonic and the micro-Brillouin light scattering techniques analysis enabled the determination of some dynamical viscoelastic parameters at higher frequencies (MHz–GHz) [5, 11], while chemical properties were followed by an original coupled micro-Raman-Brillouin and infrared spectroscopy.

2 Scaffold Elaboration and Structural Characterization

2.1 Scaffold Elaboration

Porous PCLU scaffolds were obtained through the poly(HIPE) method (High Internal Phase Emulsion Polymerized) and characterized by determination of their density, porosity and pore interconnectivity, as previously described [3]. In this study, the poly (HIPE) parameters were set to obtain PCLU scaffolds with a 85.1% porosity highly interconnected without closed voids, and a multi-scale pore sizes ranging from 600 to 1800 μm for large pore sizes, throat pore sizes as small as 150 μm and a fine porous morphology within the pore walls (pore size below 150 μm) (Fig. 1).

For control purpose, a non-porous material was prepared in the same conditions compared to the PCLU scaffold without the addition of water. The non-porous material appeared to be a clear flexible polymer.

2.2 Structural Characterization

The chemical composition of the non-porous material and the porous PCLU scaffold was monitored by Fourier-transformed infrared spectroscopy in an attenuated total reflectance mode (FTIR—ATR), as well as by μ-Raman analysis. The FTIR spectrum

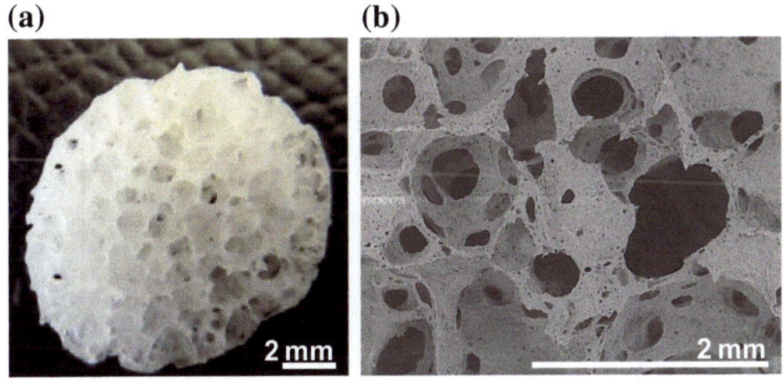

Fig. 1 Images of the porous PCLU scaffold: **a** photography; **b** environmental scanning electron microscope (ESEM—TM3000 Hitachi) image

of the non-porous material exhibits characteristic bands of poly(ester-urethane) material [3]. The porous PCLU scaffold exhibits the same characteristics bands except the presence of two bands attributed to urea –C=O groups at 1620 cm^{-1} and urea –CNH groups at 1575 cm^{-1} due to side reaction with water during the poly(HIPE) elaboration (Fig. 2a). As expected, the same bands were also found by μ-Raman analysis (Fig. 2b).

The presence of urea moieties in the porous PCLU scaffold acts as hard segments that may increase the scaffold modulus, therefore it is necessary to characterize the intrinsic elastic properties of the porous PCLU scaffold.

3 Mechanical Properties

The compressive modulus and compressive strength of the porous PCLU scaffold were measured at ambient temperature. The samples were tested in a home-made mechanical test device with a free-volume lateral expansion (uni-axial stress mode of compression). Cylindrical scaffolds of approximately 18 mm in diameter and 8 mm in height were compressed in a 500 N force range and a 7 mm displacement range at a cross-head speed of 5 mm min^{-1}. As expected, the PCLU scaffold demonstrated a typical compressive stress-strain response of an elastomeric open-cell foam (Fig. 3) [1]. At the beginning of the stress-strain curve, the stress depends linearly on the strain. The curve slope in that visco-elastic domain gives the effective modulus of elasticity E_1* of the porous PCLU scaffold. Then, the curve concaves downwards and it is followed by a long elastic collapse plateau called post-buckling domain. Above a strain of approximately 60%, the strain increases rapidly since all the pores have collapsed and the densification is reached. In the post-complete densification domain, the mechanical properties are identical to those of the solid non-porous material. The characteristic time for the slope measurement was ~10 s defining a frequency of ~0.1 Hz.

The effective modulus of elasticity E_1* of the porous PCLU scaffold was found to be 161 ± 14 kPa, which is in good agreement with the theoretical modulus calculated using Eq. 1 (E_{1theo}* = 165 kPa with N = 2.4) [7]:

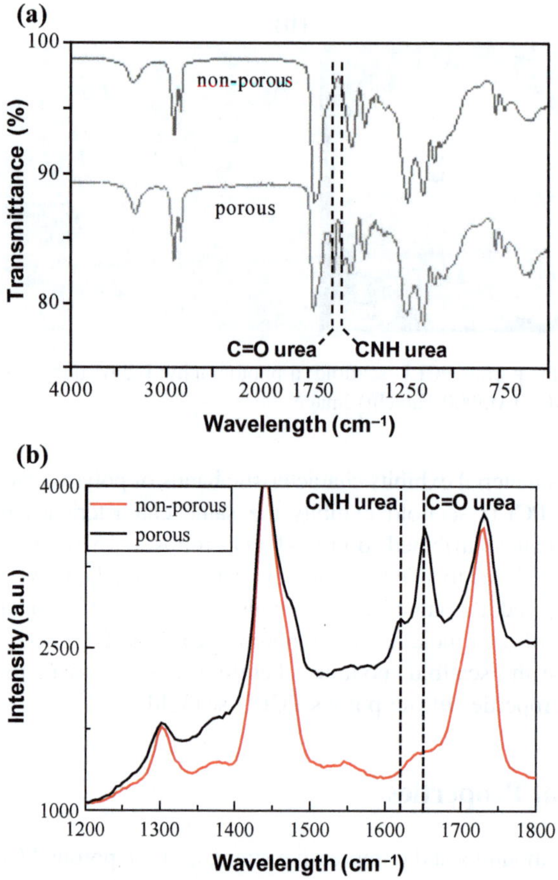

Fig. 2 Structural characterization of the non-porous material and the porous PCLU scaffold (dotted lines indicate urea functions): **a** FTIR-ATR analysis; **b** μ-Raman analysis (reduced range is shown)

$$E^*_{1theo} = E_{non-porous} \times (1 - P)^N \tag{1}$$

where $E_{non-porous}$ is the elastic modulus of the non-porous material ($E_{non-porous} = 16 \pm 2$ MPa), P is the porosity of the scaffold (85.1%), and n is an exponent value which varies between 1 and 3 depending on pore shape and orientation.

As a consequence, sufficient mechanical strength of the porous PCLU scaffold is achieved for cell culture and ultimate soft tissue formation [2].

The longitudinal sound velocity V_L was measured locally (laser spot size of a few μm^2) by the micro-Brillouin light scattering technique [5, 8] in the backscattering geometry with a $\times 100$ objective. The wavevector modulus k of the probe bulk acoustic wave is defined by $k = 4\pi n/\lambda_L$, λ_L is the incoming laser light wavelength (532 nm) and $n \sim 1.6 \pm 0.05$ the refractive index of the PCLU material. From the measurement of

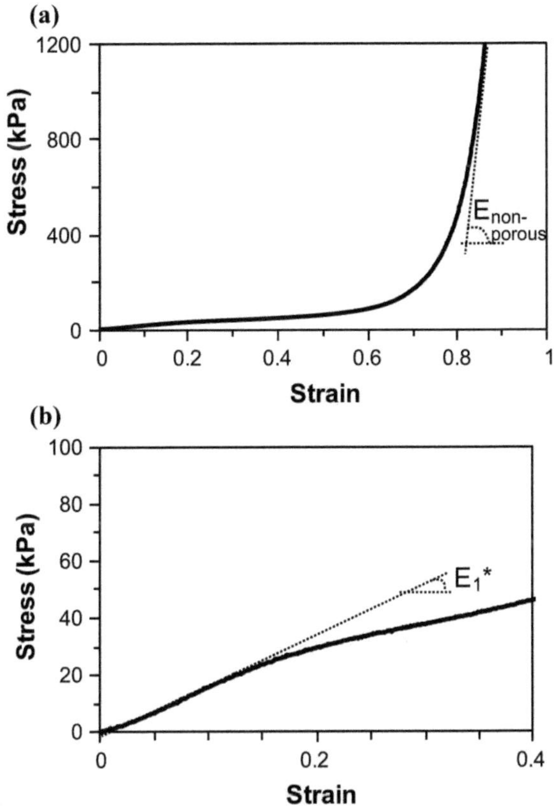

Fig. 3 Mechanical behavior of the PCLU scaffold under uni-axial stress modes of compression: **a** all strain range; **b** low strain range

the Brillouin frequency shift f_B of the inelastic scattered light, the sound velocity V_L is calculated by $V_L = f_B\lambda_L/2n$. A typical spectrum is shown in Fig. 4 for the non-porous material and the porous PCLU scaffold. The Brillouin frequency shift lies close to 13 GHz while their full linewidth at half-maximum are $\Gamma_{porous} = 2140$ MHz and $\Gamma_{non\text{-}porous} = 1873$ MHz, respectively. We measured $V_L = 2160 \pm 110$ m/s and the related longitudinal elastic constant C = 4.8 ± 0.4 GPa. The dynamical longitudinal viscosity ($\eta = \rho\Gamma_d/k^2$) could be estimated from the natural linewidth (Γ_d) obtained after deconvolution of the instrument linewidth (418 MHz), $\eta \sim 1.3$ mPa s.

These high-frequency results are higher than the ones measured by the ultrasonic pulse-echo method with a 10 MHz longitudinal transducer, $V_L = 1773 \pm 53$ m/s and C = 3.2 ± 0.15 GPa.

Considering a Poisson ration $\nu = 0.425$, we calculated the Young modulus for each frequency 10 MHz (1203 MPa) and 13 GHz (1788 MPa) and plotted them with the quasi-static result (16 MPa) in a log-log scale as a function of the frequency, in Fig. 5. Finite relaxation time and low compressibility provide a qualitative explanation for the

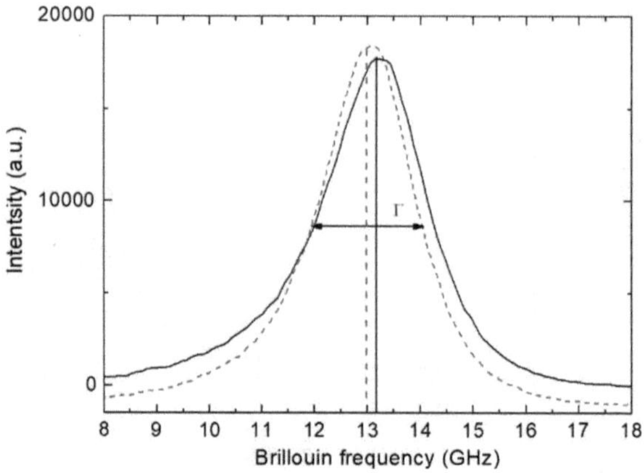

Fig. 4 μ-Brillouin light scattering spectra of the non-porous (dashed line) material and the porous (line) PCLU scaffold. The peak is the inelastic scattering of light by the longitudinal bulk acoustic wave with a frequency shift $f_B \sim 13$ GHz

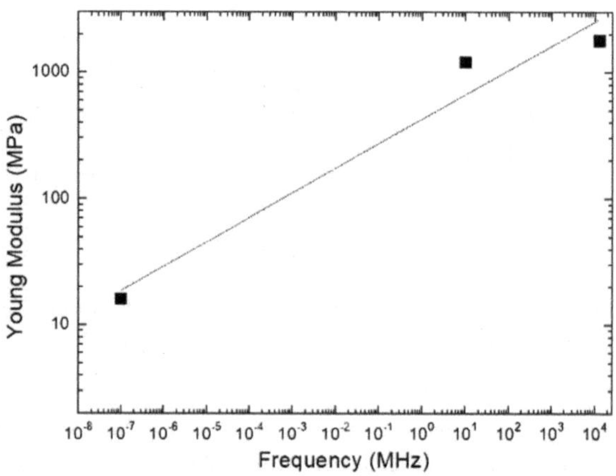

Fig. 5 Young modulus as a function of the frequency (log-log scale) of the PCLU non-porous polymer. The dotted line is log-log linear fit

observed large difference in modulus between the ultrasound, Brillouin light scattering and standard quasi-static mechanical tests.

Recent rheological studies have shown that the mechanical modulus (M) of many soft materials follows a power-law dependence on frequency (f):

$$M = M_0 \times \left(\frac{f}{f_0}\right)^{\alpha} \tag{2}$$

Here, M_0 and f_0 are scale factors for stiffness and frequency, respectively, and α is the scaling exponent factor ($\alpha = 0$ for purely elastic and $0 < \alpha < 1$ for visco-elastic materials) [6, 9]. From the best-fit in Fig. 5, we found for the Young modulus $\alpha = 0.194$ and $\log(E_0/(f_0)^{\alpha}) = 2.634$. We need supplementary information on the loss-modulus or on the dynamical shear modulus G to determine separately, the scale factors E_0 and f_0 [8].

4 Conclusion

Quasi-static uni-axial tests provided the effective Young modulus of porous PCLU (165 kPa) biopolymer and of the dense one, achieving sufficient strength for cell culture and ultimate soft tissue formation. Our measurements revealed a strong linear correlation (log-log scale) between hypersonic (MHz–GHz) and low-frequency Young modulus. It is attributed to the power-law scaling of modulus in frequency. The power-law dependence in the Young modulus has been measured at frequencies up to $\sim$10 GHz.

Acknowledgements. The authors would like to thank the Interdisciplinary Institute of Experimental Sciences of Université Paris 13 for the support and allocated grants.

References

1. Ben-Dor G, Mazor G, Cederbaum G, Igra O (1996) Stress-strain relations for elastomeric foams in uni-, bi- and tri-axial compression modes. Arch Appl Mech 66:409–418. https://doi.org/10.1007/BF00803675
2. Burdick JA, Mauck RL (2011) Biomaterials for tissue engineering applications. Springer, Wien
3. Changotade S, Radu-Bostan G, Consalus A, Poirier F, Peltzer J, Lataillade J-J, Lutomski D, Rohman G (2015) Preliminary in vitro assessment of stem cell compatibility with cross-linked poly(ε-caprolactone urethane) scaffolds designed through high internal phase emulsions. Stem Cells Int 2015:283796. https://doi.org/10.1155/2015/283796
4. Chen Q, Liang S, Thouas GA (2013) Elastomeric biomaterials for tissue engineering. Progr Polym Sci 38:584–671. https://doi.org/10.1016/j.progpolymsci.2012.05.003
5. Djemia P (2009) La diffusion Brillouin pour caractériser les propriétés élastiques ou magnétiques de multicouches. In: Elmazria O (ed) Les ondes en instrumentation, Instrumentation, Mesure, Métrologie. Hermès Science publications – Lavoisier, Cachan
6. Fabry B, Maksym GN, Butler JP, Glogauer M, Navajas D, Fredberg JJ (2001) Scaling the microrheology of living cells. Phys Rev Lett 87:148102. https://doi.org/10.1103/PhysRevLett.87.148102
7. Hou Q, Grijpma DW, Feijen J (2003) Porous polymeric structures for tissue engineering prepared by a coagulation, compression moulding and salt leaching technique. Biomaterials 24:1937–1947. https://doi.org/10.1016/S0142-9612(02)00562-8
8. Scarcelli G, Kim P, Yun SH (2011) In vivo measurement of age-related stiffening in the crystalline lens by Brillouin optical microscopy. Biophys J 101(6):1539–1545. https://doi.org/10.1016/j.bpj.2011.08.008
9. Sollich P, Lequeux F, Hébraud P, Cates ME (1997) Rheology of soft glassy materials. Phys Rev Lett 78:2020–2023. https://doi.org/10.1103/PhysRevLett.78.2020

10. Tee S-Y, Fu J, Chen CS, Janmey PA (2011) Cell shape and substrate rigidity both regulate cell stiffness. J Biophys 100:L25–L27. https://doi.org/10.1016/j.bpj.2010.12.3744
11. Zhang J, Zhang H, Wu L, Ding J (2006) Fabrication of three dimensional polymeric scaffolds with spherical pores. J Mater Sci 41:1725–1731. https://doi.org/10.1007/s10853-006-2873-7

Study of Al 2017 Alloy Prepared by Recycling Method via Powder Metallurgy Route

Mariem Bhouri[1(✉)] and Foued Mzali[1,2]

[1] Laboratory of Thermal and Energetic Systems Studies, National Engineering School of Monastir (ENIM), University of Monastir, Monastir, Tunisia
bhourimaryem@gmail.com, foued.mzali@enim.rnu.tn
[2] Department of Mechanical Engineering, National Engineering School of Monastir (ENIM), University of Monastir, Monastir, Tunisia

Abstract. Recycling of Al 2017 alloy machining chips via powder metallurgy route including hot compaction and sintering processes was studied. This method was proposed in order to elaborate Aluminium products based on Al 2017 alloy powder. To obtain desired chips with small thickness, the optimal cutting conditions are applied. The chips obtained after turning operation had undergone a milling step by varying milling times. Then, these powders were hot compacted and then sintered at various temperatures and times. The results indicated a decrease of particle size of 2017A alloy by increasing the milling time which reaches 70 µm after 20 h of milling. This is due to the successive plastic deformation which makes particles hard and brittle. By using DSC analysis, a high level of stored enthalpy release is observed due to the milling process where two kinds of structural changes are presented. Also, heating the deformed powder provides the elimination or the rearrangement of some dislocations and the release of the residual stresses stored in the powder. Then, it can be deduced that higher densities are obtained at higher sintering parameters (t2 = 90 min, T2 = 550 °C) which explained by particle size refinement during the milling step and the activated phenomenon of diffusion during the sintering step.

Keywords: Al 2017 alloy chips · Aluminium recycling · Powder metallurgy route · Microstructure · Density

1 Introduction

The use of primary Aluminium to manufacture products requires a lot of energy. This is why recycling methods of Aluminum and its alloy in the form of chips becomes a major issue. These chips can be recycled by conventional methods by involving re-melting method. However, it is difficult due to their elongated spiral shape and small size, also, the oil oxides were coated on the surface of the chip which makes the remelted alloys not good to use. Then, it was lead to high pollution that mainly due to the fumes and dross generated during re-melting [1]. An average between 30% and 46% of the metal is lost during the re-melting process [2, 3]. To avoid these disadvantages, direct conversion method of recycling Aluminium chip has been introduced without introducing the re-melting process which was based on powder metallurgy

F. Chaari et al. (eds.), *Advances in Materials, Mechanics and Manufacturing*, Lecture Notes in Mechanical Engineering,
https://doi.org/10.1007/978-3-030-24247-3_2

technique. This method was compared with the conventional method and it has been proved that the direct conversion method is more suitable in terms of energy consumption and material saving [1]. Then, this method was used where the Al 6060 alloy chips reinforced with Al_2O_3 [4] or SiC [5] are directly cold pressed and then hot extruded. As results, they concluded that these products can lead to similar or high mechanical properties than the use of cast Al 6060 alloy. Gronostajski et al. [6] mixed recycled Aluminium powder with reinforcing phase to produce composite via cold compaction and hot extrusion and the characterization of their mechanical and tribological properties showed that the use of powder as the initial material improve the mechanical behavior of composites. Chiba and Yoshimura [7] used the hot extrusion step to recycle cast Al–Si alloy chips and then they concluded that a higher ductility is shown for recycled material with a reduction of around 30% in the ultimate tensile strength compared to the primary alloy.

The possibility of recycling Al 2017 alloy machining chips obtained by lathe turning has not been studied yet which is the aim of this study. The microstructure, such as size, shape, and morphology of Al 2017 alloy powder during milling step with varying milling time are examined. Then, the effect of factors related to the sintering processes such as sintering temperature and time on recycled Al 2017 alloy are also studied. The density of different specimens of recycled Al 2017 alloy obtained under different milling time and cycling conditions of sintering are compared.

2 Experimental Procedure and Characterization of Mechanical Properties

Al 2017 alloy chips were used as the starting materials. Table 1 presented the basic elements such as Cu, Mg, and Si of the Al 2017 alloy. To obtain desired chips with a small thickness, the cutting conditions, namely, cutting rate, feed rate, and cut depth, are adopted during the turning operations were 200 m/min, 0.2 m/min, and 0.25 mm.

Table 1 The chemical composition of the Al 2017 alloy

Al	Si	Fe	Cu	Mn	Mg
Bal.	0.66	0.27	4.01	0.64	0.57

Then, the cutting process is exploited to reduce the size of chips. This latter will be passed by milling step in order to obtain a powder with a different size as a function of milling time with using of 0.3 ml methanol and (10:1) ball to powder weight ratio to reach a fine powder without his stick to the ball and the container wall. After, this powder will be hot pressed by applying 150 MPa at 200 °C for 2 h. Then, the hot pressed samples were sintered in a furnace for different combinations with 20 °C/min of heating. The sintering temperature and time is about 450 °C (550 °C) and 45 min (90 min).

To evaluate the effect of the milling process, particle size and morphology of recycled Al 2017 alloy powder and their sintered were determined using an optical microscope and an image processing program «ImageJ». Differential scanning calorimeter from room temperature to 550 °C was performed at a rate of 5 °C/min under a pure nitrogen atmosphere in order to determine the structural changes in the powder during the annealing process. Also, their densities were measured based on the measurement of dimensions and weight of each sample.

3 Results and Discussion

3.1 Particle Size

Figure 1 presents OM micrographs showing the morphology of milled Al 2017 alloy powder for different times of milling. It can be noticed that the particle size of milled powder decreases with the increase of milling time. Because of a good ductility of aluminum and its alloys, particularly our alloy, for short milling time (5 h), the ductile milled Al 2017 alloy powder tend to deform and weld cold. These mechanisms are proved by the increase in their size compared to unmilled particles. But with long milling times (after 7.5 h), there is a decrease in the particle size. This is due to

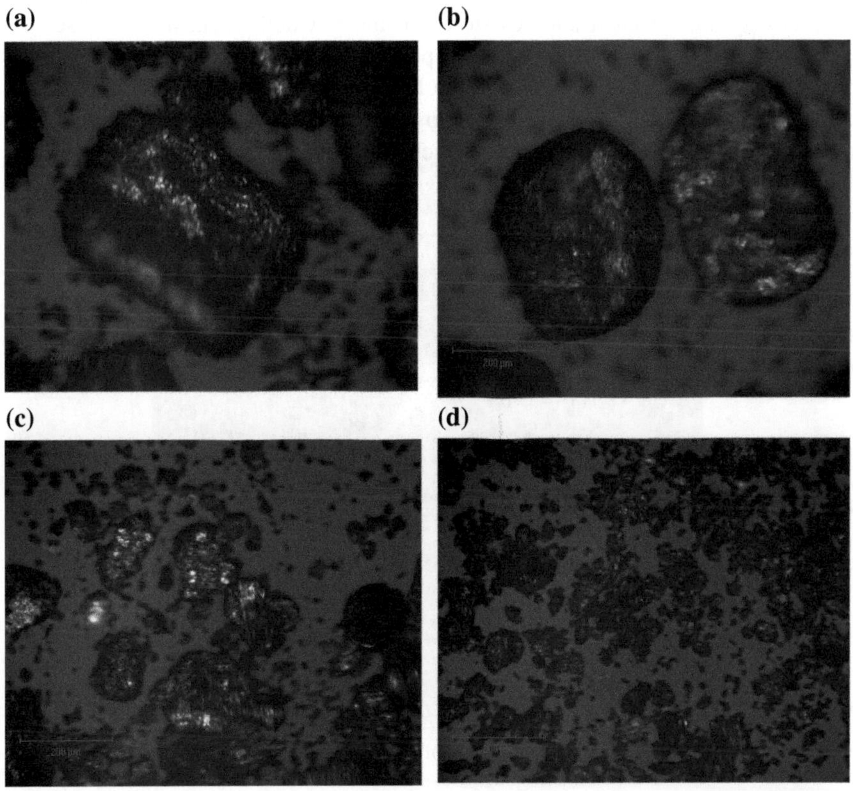

Fig. 1 OM micrographs of powder **a** unmilled, **b** 5 h, **c** 7.5 h, **d** 20 h

successive plastic deformation which makes particles hard and brittle and the fracture mechanism will be active. Zhao et al. [8] confirmed these results when they used a milled Al 6061 alloy particles. And for a long period of milling, Ramezani and Neitzert [9] proved that the fracture mechanism is dominating.

An image processing program «ImageJ» is used to determine the particle size of milled particles. This method is used by Khorasani et al. [10]. The determination of Feret Diameter (Fig. 2) indicates the average size of particles after different times of milling. Table 2 presented a decrease in particle size from 450 μm after 2.5 h to 70 μm after 20 h of milling.

Figure 3 illustrates the DSC analysis for the powders milled for different periods showing a specific heat (Cp) presented a quantity of energy released during the annealing step. An increase in their values with the increase of milling time. This is due to the increase in the density of dislocations according to successive plastic deformation. A wide exothermic regime is observed from 100 to 450 °C. This regime may be attributed to the presence of the precipitation Al_2Cu into Al 2017 alloy. According to binary phase diagram (Al–Cu), this sequence with their structural change provokes an important deformation of the matrix and then, during the annealing step, this deformation is released in the form of a huge quantity of enthalpy.

In addition, Révész and Lendvai [11] and Abdoli et al. [12] explained that heating the deformed powder provides the required activation energy for releasing part of the elastic energy and residual stresses stored in the powder. Then, the increase of the heating process exposed by an exothermic peak at 500 °C contributes to the generation of the recrystallization process driving the formation of new grains and a decrease of the free energy. These mechanisms are proved by Tellkamp et al. [13]. After 500 °C, there is a decrease in the Cp values which can be explained by the dissolution of

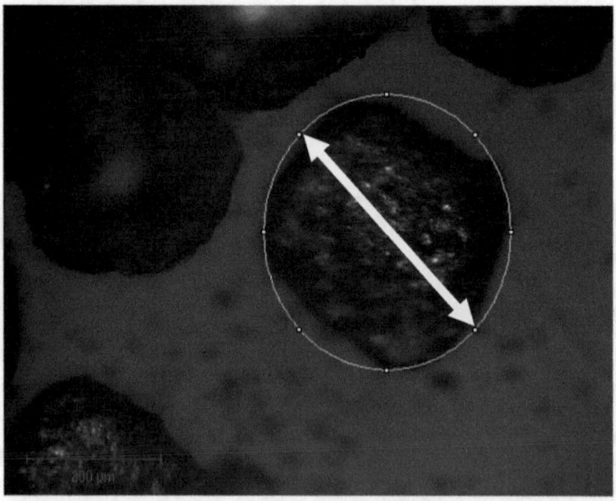

Fig. 2 Schematic drawing of Feret diameter for a particle

Table 2 The particle size of raw and milled powder at different milling times

Milling time (h)	0	2.5	5	7.5	10	15	20
Feret diameter (mm)	370.3	450.4	310.9	258.7	104.1	100.4	~70

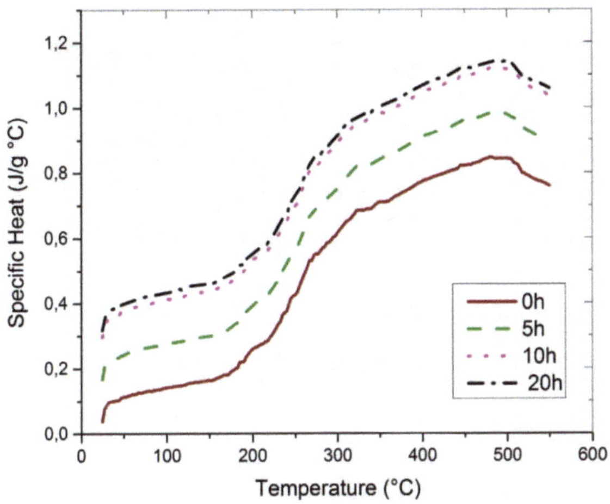

Fig. 3 DSC analysis of milled powders after various milling times

precipitate phases and the diminution of lattice distortion induce then a decrease of the Cp values.

3.2 Microstructure of Sintered 2017A Alloy

Characteristic OM micrographs of Al 2017 alloy with different milling times and different combinations during sintering are shown in Fig. 5.

The microstructure of the metal phase (bright) shows a good chemical bonding among the Al 2017 alloy particles in result having a denser structure as seen in Fig. 4. It can see a visible residual porosity (black regions) which disappears with the increase of the sintering parameters and consequently, the formation of a denser structure as a result of the higher diffusion rates. Thus, the sintering temperature plays an important role as controlling parameter in the sintering process.

For the effect of sintering time as seen in Fig. 4a–c, it can conclude that for 450 °C there is less porosity for 90 min of sintering than 45 min. This explained that particles of metal do not have much time to unite and to create a denser structure.

3.3 Density

The variation of the sintered porosity for different sintering temperatures is exposed in Fig. 5. Four sintering curves proved that the sintered porosity gradually decreases with the increase of the milling period. This can be explained by the adapted morphology of particles after the milling process. According to Fig. 1, it is clear that the morphology

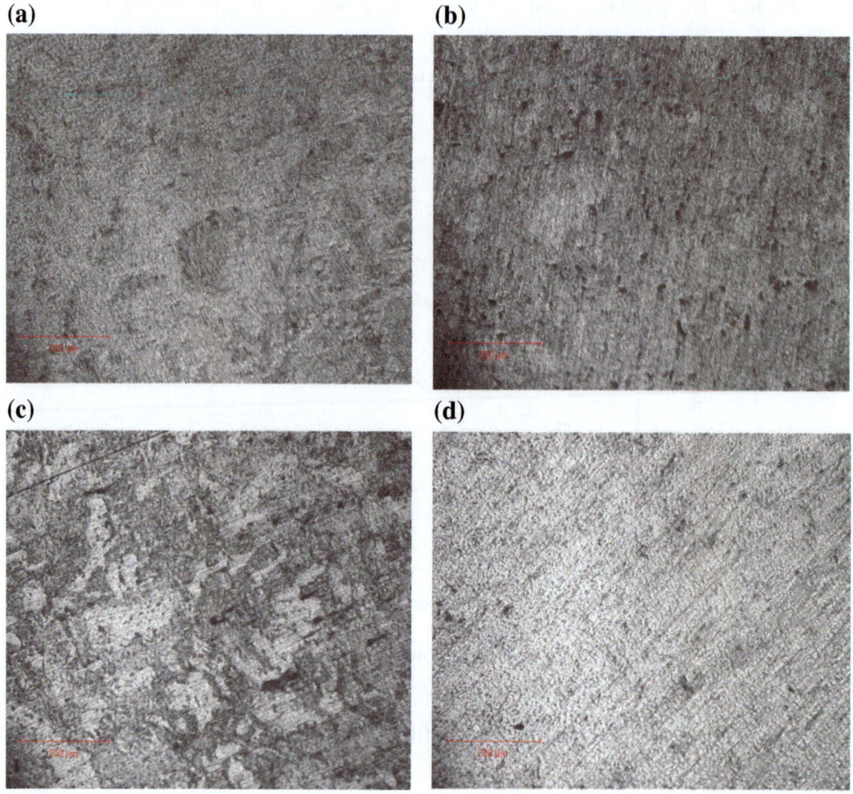

Fig. 4 OM images of sintered Al 2017 alloy for milled powder for 20 h at different sintering times and temperatures; **a** [t1 T1], **b** [t1 T2], **c** [t2 T1], **d** [t2 T2]

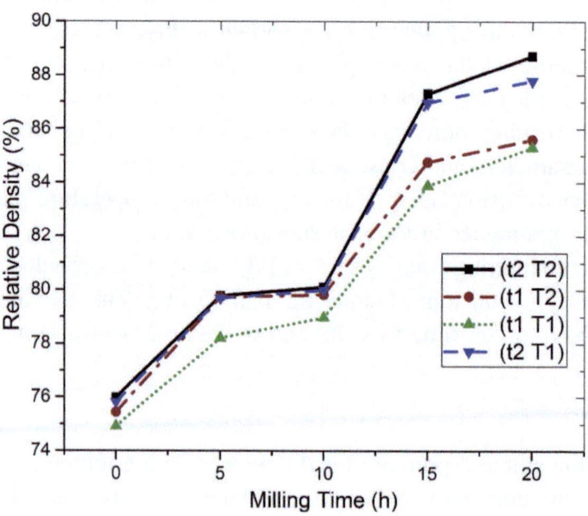

Fig. 5 Variation of porosity as a function of milling time, sintering time and sintering temperature

of the particles is flattened for a short period of milling. But with the increase of milling time, the fracture mechanism will be active and particles change their morphology to become equiaxed. According to Fogagnolo et al. [14], these morphologies affect in the powder packing during compaction step and consequently interfere in the sintered density of the powder. Then according to these references [15,16], with the increase of sintering parameters, there is an increase of the relative density and as a result, the formation of denser structure with some porosity.

We can also determine that the effect of the sintering temperature is more significant than the sintering time on porosity. As a result, the microstructure is denser after the optimal sintering combination (90 min, 550 °C).

4 Conclusion

In the current study, the feasibility of recycling Al 2017 alloy via powder metallurgy route was investigated, by using a direct conversion method. A decrease of particle size of Al 2017 alloy by increasing the milling time is observed. This phenomenon is proved by the successive plastic deformation which makes particles hard and brittle. Particles are characterized by two kinds of structural changes during annealing which explained by a high level of stored enthalpy is observed due to the milling process. Then, a denser microstructure is obtained at higher sintering parameters which explained by particle size refinement during the milling step and diffusion phenomenon during the sintering step.

References

1. Samuel M (2003) A new technique for recycling aluminum scrap. J Mater Process Technol
2. Sherafat Z, Paydar MH, Ebrahimi R (2009) Fabrication of Al7075/Al, two-phase material, by recycling Al7075 alloy chips using powder metallurgy route. J Alloy Compd 487:395–399
3. Gronostajski J, Marciniak H, Matuszak A (2000) New methods of aluminum and aluminum-alloy chips recycling. J Mater Process Technol 106:34–39
4. Fogagnolo JB, Ruiz-Navas EM, Robert MH, Torralba JM (2003) The effects of mechanical alloying on the compressibility of aluminum matrix composite powder. J Mater Sci Eng A355:50–55
5. Tekkaya AE, Schikorra M, Biermann D, Hammer N, Pantke K (2009) Hot profile extrusion of AA-6060 aluminium chips. J Mater Process Technol 209:3343–3350
6. Gronostajski J, Matuszak A (1999) The recycling of metals by plastic deformation: an example of recycling of aluminum and its alloys chips. J Mater Process Technol 92–93:35–41
7. Chiba R, Yoshimura M (2015) Solid-state recycling of aluminum alloy swarf into c-channel by hot extrusion. J Manuf Process 17:1–8
8. Zhao N, Nash P, Yang X (2005) The effect of mechanical alloying on SiC distribution and the properties of 6061 aluminum composite. J Mater Process Technol 170(3):586–592
9. Ramezani M, Neitzert T (2012) Mechanical milling of aluminum powder using planetary ball milling process. J Achievements Mater Manuf Eng 55(2):790–798

10. Khorasani S, Abdizadeh H, Heshmati-Manesh S (2014) Evaluation of structure and morphology of aluminum powder particles milled at different conditions. Adv Powder Technol 25(2):599–603
11. Révész Á, Lendvai J (1998) Thermal properties of ball-milled nanocrystalline Fe, Co, and Cr powders. Nanostruct Mater 10(1):13–24
12. Abdoli H, Ghanbari M, Baghshahi S (2011) Thermal stability of nanostructured aluminum powder synthesized by high-energy milling. J Mater Sci Eng A528:6702–6707
13. Tellkamp VL, Dallek S, Cheng D, Lavernia EL (2001) Grain growth behavior of a nanostructured 5083 Al-Mg alloy. J Mater Res 16(4):938–944
14. Fogagnolo JB, Ruiz-Navas EM, Robert MH, Torralba JM (2003) The effects of mechanical alloying on the compressibility of aluminium matrix composite powder. Mater Sci Eng, A 355(1–2):50–55
15. Rahimian M, Ehsani N, Parvin N, Baharvandi HR (2009) The effect of particle size, sintering temperature and sintering time on the properties of Al-Al_2O_3 composites, made by powder metallurgy. J Mater Process Technol 209(14):5387–5393
16. Latief FH, Sherif ESM (2012) Effects of sintering temperature and graphite addition on the mechanical properties of aluminum. J Ind Eng Chem 18(6):2129–2134

Initiation Life Prediction Method for Defective Materials

Maroua Saggar[1(✉)], Anouar Nasr[2], and Chokri Bouraoui[1]

[1] Laboratoire mécanique de Sousse, Ecole nationale d'Ingénieurs de Sousse, Université de Sousse, Bp.264 Erriadh, 4023 Sousse, Tunisia
{marwasaggar, chokribouraoui2}@gmail.com

[2] LGM, Institue préparatoire aux études d'Ingénieurs de Monastir, Université de Monastir, Monastir, Tunisia
Anoir_nasr@yahoo.fr

Abstract. The objective of the present paper emphasizes the investigation of the fatigue strength of materials containing surface defects and the development of a comprehensive analytical model, based on the response surface methodology (RSM), for correlating the interactive and higher-order influences of the various parameters such as the size of the defect, loading, and load ratio on the residual lifetime of these components. Firstly, a 3D-finite element analysis of a specimen containing a surface defect using ABAQUS commercial software is established to (i) simulate the distribution of the stress and the equivalent plastic deformation at the vicinity of the defect and (ii) to predict the fatigue life using the Smith-Watson-Topper **SWT** model. The non-linear kinematic hardening model, coupled with Lemaitre and Chaboche's damage model, is used to characterize the material behavior. For this work, we used the experimental design technique to characterize the effects and the interactions between the defect size, the loading and the load ratio given by the Smith Watson Tooper model. Finally, using the surface response method, we have been able to develop an analytical model capable of predicting the fatigue response of steel 1045 considering defect size, loading and load ratio. Consequently, the finding results show a good agreement with those experimentally obtained.

Keywords: Defect · Crack initiation · Smith watson tooper · Experimental design technique

1 Introduction

The damage comes from the cyclic nature of the stresses applied to the mechanical parts. The fatigue phenomenon is the most detrimental failure mechanism of mechanical structures in several industrial sectors such as automotive, marine and aeronautics [1]. During cyclic loading, micro cracks can be initiated and propagated in mechanical structures, leading to an unexpected failure, even when the ultimate strength of the material is much higher than the applied stress level [2]. When these cracks occur, the remaining lifetimes of these components are of paramount importance in the industry to properly ensure the safety behavior of these components. Since the defects are inevitable during the machining process [3, 4], manufacturing companies

F. Chaari et al. (eds.), *Advances in Materials, Mechanics and Manufacturing*, Lecture Notes in Mechanical Engineering,
https://doi.org/10.1007/978-3-030-24247-3_3

always try to reduce the risk of ruin parts during operation in case of defective materials and to determine the maintenance methods in order to change the damaged parts before their loss. These defects are very damaging to fatigue resistance and generally lead to a considerable reduction in their serviceability [5–7].

Hence, it is very important to find the accurate method to predict the fatigue crack propagation in a structural component containing defect before final failure. In the present study, we will propose an approach based on the Smith Watson Tooper method SWT [8] to predict the onset of crack initiation in defective materials and the surface response method to study the effects of the main parameters (loading and defect size) influencing the polycyclic fatigue life of these defectives' specimens.

Investigations are made for defective 1045 steel subjected to tension loading, by corresponding numerical simulation. Comparisons have also been made to analyze the validity of the improved model. A significant phase shift between numerical values and experimentation was detected. Finally, using the experimental design technique we evaluated the contributions of defect size, loading and load ratio over fatigue life and we have been able to develop an analytical model capable of predicting the fatigue response of steel 1045. The developed model has been validated in the case of 300 μm defects of spherical shapes.

2 Finite Element Modeling

A single surface defect specimen containing a stationary defect size $\sqrt{\text{area}} = 300$

μms considered to investigate the correct mechanical response "cyclic plastic behavior" in the vicinity of each defect.

The EDM [9] is the process chosen to introduce artificial surface defects on the test sample. This method allows a good reproducibility of the geometries and sizes of defects. In order to optimize the computation time, only a small volume surrounding the defect is modeled as shown in Fig. 1. Considering loading and geometry symmetries, the completed model is simplified. Only a quarter of this small volume is considered for the FE-analysis. The validation of these simplifications has been verified in the work of Gadouini [9]. The choice of the cube (for tension) is based on the experimental observation showing that the crack propagates in the plane perpendicular to the maximum principal stress in opening mode I [10].

Due to the fact that cracks start to propagate from the surface of specimen [10], a very fine mesh of type (C3D4R) is adopted for the SENT specimen and implemented in the region of defect as illustrated in Fig. 2. In order to consider, the mean stress relaxation, the Bauschinger effect and the cyclic hardening during cyclic loading, Lemaitre and Chaboche kinematic hardening model, embedded in the commercial code ABAQUS, is used.

The law of the elastoplastic behavior with nonlinear kinematic hardening proposed by Chaboche [11] used to describe the plasticity in the 1045 steel is given by the following form:

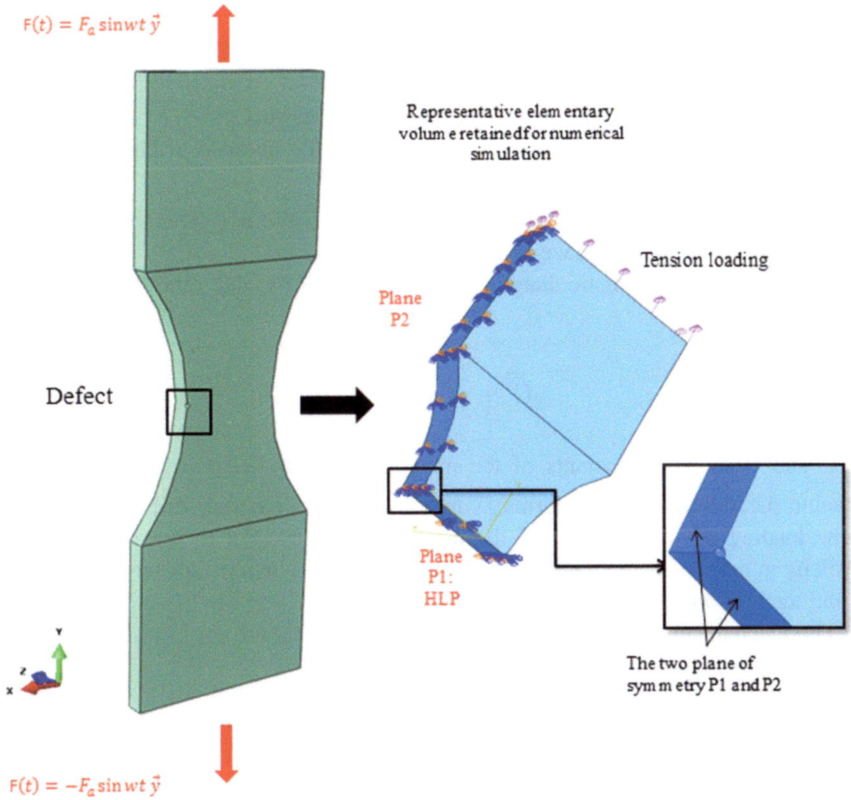

Fig. 1 Fully completed model and simplification of the FE calculation model for a spherical defect with applied load and boundary conditions

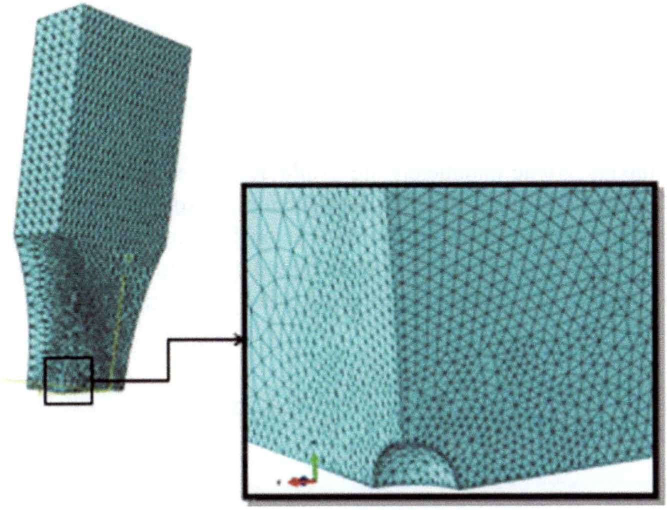

Fig. 2 Finite element mesh of the defective specimen

$$f = j_2(\overline{\overline{\sigma}} - \overline{\overline{X}}) - k \tag{1}$$

where

J_2 is the second invariant of the deviatoric stress tensors.

k is the radius of the cylinder of Von Mises in the absence of hardening (initial radius).

$\overline{\overline{X}}$ is the back stress tensor associated to kinematic hardening. This tensor represents the position of the yield surface in the stress space.

The nonlinear kinematic hardening is described by the following differential equation:

$$d\overline{\overline{X}} = \frac{2}{3} C d\varepsilon^p - \gamma \overline{\overline{X}} dp \tag{2}$$

C and γ are two constants of the material. The nonlinearity introduced in this equation (i.e. described by the terms $\gamma \overline{\overline{X}} dp$) makes it possible to describe the effects of a cyclic loading on the behavior of the material [11]. In order to use the criterion of plasticity in numerical simulation, the identification of its own parameters is determined by the method indicated in the Abaqus documentation.

The mechanical and cyclic parameters of the material are given in Table 1.

Table 1 Mechanical and cyclic parameters of the 1045 steel

Young modulus (MPa)	205,000
Yield stress (MPa)	580
Poisson's ratio	0.3
Q	269
C	33,000
γ	124

3 SWT Model's Capability for Predicting Fatigue Life

Several multiaxial fatigue parameters are available to predict the onset of crack initiation in mechanical structures. The fatigue behavior of a specimen containing defects is studied using the empirical model of Smith Watson Tooper [8]. The damage parameter proposed by Smith et al. is commonly accepted for engineering applications. The sample is subjected to a cyclic fatigue load with a load ratio R = −1 for different applied loads.

$$\mathrm{DP_{SWT}} = \sigma_{\max} \frac{\Delta\varepsilon_{\max}}{2} = \frac{(\sigma'_f)^2}{E} (2N_i)^{2b} + \varepsilon'_f \sigma'_f (2N_i)^{b+c} \tag{3}$$

where Ni, σ_{max}, $\Delta\varepsilon max2$, are the number of cycles, the maximum normal stress, the total strain amplitude and the modulus of elasticity, respectively. $\varepsilon f'$ and present the low-cycle fatigue coefficients, $\sigma f'$ and are the high-cycle fatigue constants.

The SWT model (Eq. 3) is implemented to predict the fatigue life of defective materials. The fatigue parameters of SWT for steel 1045 are shown in Table 2 for purely alternating tension and torsion.

Table 2 1045 strain data life

Fatigue strength coefficient in tension $\sigma_f^{'}$	998 MPa
Fatigue strength coefficient in torsion $\tau_f^{'}$	471 MPa
Fatigue strength exponent b	−0.095
Shear modulus G	78.9 GPa
Young modulus E	205 GPa

Figure 3 shows a good agreement between the experimental points [7] and the SWT model for predicting fatigue life to predict the onset of crack initiation in defective materials.

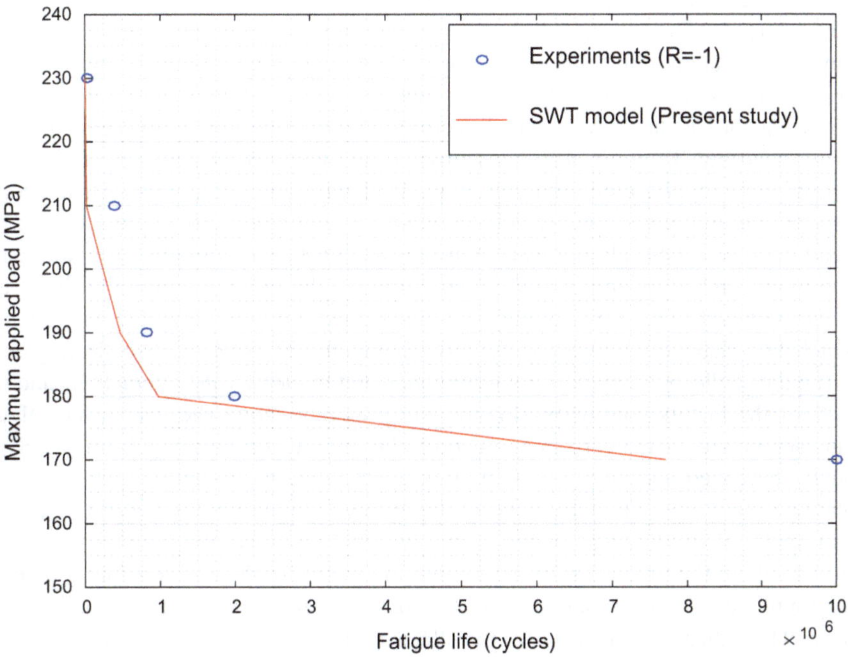

Fig. 3 Comparison between experimental results and simulation results

4 Response Surface Methodology

The response surface design methodology is a very interesting tool for building empirical models when the true linkage function between the process parameters and the desiring response is very complicated.

In this study, three input parameters: external load, load ratio and defect size are chosen as the main factors for generating the factorial experimental design and their respective levels in actual and coded values are shown in Table 3.

Table 3 Controllable factors adopted for testing

Factors	Notation	Level		
		−1	0	1
$\sqrt{area}$ (μm)	A	300	400	500
σ_A (MPa)	B	210	220	230
Load ratio	C	−1	0.1	0.5

To do this, a methodology based on experimental plans is putted in place. Then, in order to optimize the performance of the simulations, we will study the sensitivity of the various factors to the variability of our forecast.

Experiment plans help to understand how a system responds to parameters that might change it. They present an interesting tool for the designers to have the maximum of information of the effect of variation of the various parameters of entries on the behavior of a system with a minimum of test.

Each factor has three levels, which requires 27 data collection experiments. Fatigue life is considered the main output response. In the first step, we define the variation range of each input factors: external load, load ratio and defect size. Then, model geometry, material parameters (1045 steel) and the loading conditions are fixed. In the second step, an elastic plastic analysis using the non-linear isotropic/kinematic hardening model embedded in ABAQUS is performed. The residual lifetime of the structure is predicted. The SWT model is implemented for predicting the fatigue life. Minitab 18.0 statistical software is used to generate the 27 experiments collecting data.

The data collection experiments were conducted to identify the interactions and the effects of the factors considered. Figure 4 respectively shows the main effect of the input factors and their interactions on both the fatigue life resulting from the SWT model. The coded variables (−1, 0, 1) that characterize the levels of each factor are plotted on the x-axis. The y-axis shows the average fatigue life for a given level.

From the effect diagrams in Fig. 4, we find that the three parameters have significant effects on the prediction of the fatigue life of the 1045 steel containing defects. The graph allows us to visualize the effects of each parameter on the response of our system. These diagrams characterize the effects of the three parameters ($\sqrt{area}$, applied loading and load ratio) on the output (fatigue life at initiation). The level of each input parameter is represented on the x-axis of the graph. The y-axis represents the value of average fatigue life given for each level.

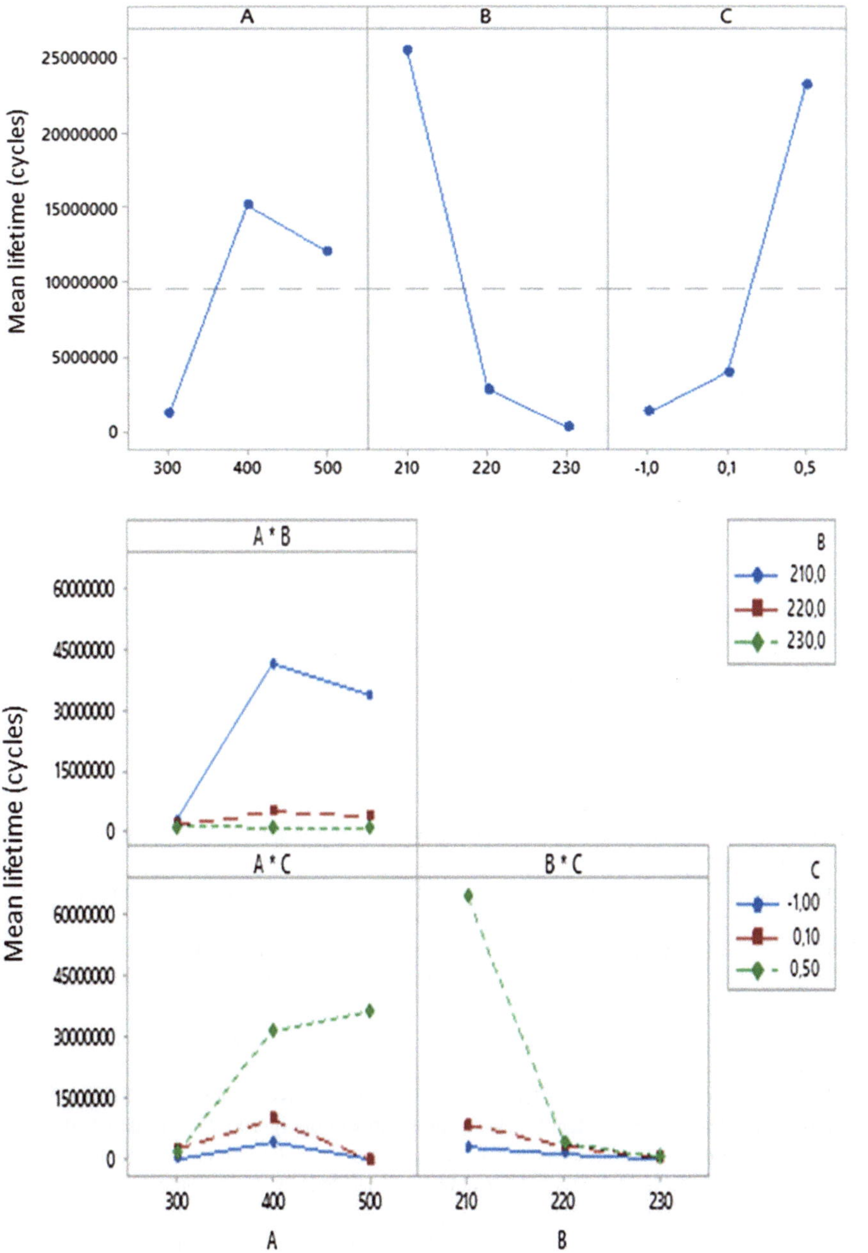

Fig. 4 The representation of the main effects and interactions

The experimental design technique allowed us to develop a simple analytical model capable of expressing the input factors of the system (defect size, loading and load ratio) as a function of the output (fatigue life). The proposed model is expressed as follows:

$$\begin{aligned} \text{Fatigue life} &= 4{,}325{,}627{,}079 + 2{,}465{,}551\,\text{A} - 42{,}449{,}602\,\text{B} \\ &\quad + 362{,}184{,}621\,\text{C} - 847\,\text{A2} + 100{,}228\,\text{B2} \\ &\quad + 30{,}284{,}354\,\text{C2} - 7826\,\text{A}^*\text{B} + 89{,}997\,\text{A}^*\text{C} - 1{,}675{,}073\,\text{B} * \text{C} \\ \text{R}^2 &= 97.93\% \end{aligned} \tag{4}$$

5 Conclusion

The application is established to validate the SWT model's effectiveness for predicting the fatigue life of defective materials. Therefore, a defective specimen is considered in this study. The specimen was applied to a cyclic fatigue loading with load ratio (R = −1), for different applied loads. The SWT model is implemented to predict the fatigue life at the initiation of the crack. Numerical results were compared with experimental results and a reasonably good agreement was observed. It has been proven that all these parameters are involved in the fatigue behavior of the steel 1045 with the major role of the load ratio factor.

Acknowledgements. This work is partially supported by the Laboratory of Mechanic of Sousse. The authors also gratefully acknowledge the helpful comments and suggestions of the reviewers, which have improved the presentation.

References

1. El-Hadek MA (2014) Fatigue characterization of fluorogold and fluorogreen polymers. Metall Mater Trans A 45(1):317–323
2. Billaudeau T (2002) Fatigue multiaxiale des matériaux à défauts: mécanisme et critère d'endurance, thèse doctorale, ENSMA, Poitiers
3. Crossland B (1956) Effect of large hydrostatic pressures on the torsional fatigue strength of an alloy steel. In: The international conference on fatigue of metals. Institution of mechanical engineers, London, pp 138–149
4. Murakami Y (2002) Metal fatigue: effects of small defects and nonmetallic inclusions. Elsevier
5. Nasr A, Hassine W, Bouraoui CH (2017) Fatigue limit assessment for defective materials based on affected depth. Metall Res Technol. https://doi.org/10.1051/metal/2017045
6. Saggar M, Sallem H, Bouraoui C (2017) Fatigue life prediction under variable loading based on a new damage model devoted for defective material. Int J Adv Manuf Technol. https://doi.org/10.1007/s00170-017-1198-9
7. Saggar M, Bouraoui C, Nasr A (2018) Fatigue reliability prediction of defective materials based on a useful equivalent Wöhler curve. Int J Adv Manuf Technol. https://doi.org/10.1007/s00170-018-1975-0
8. Smith KN, Watson P, Topper TH (1970) A stress-strain function for the fatigue of metals. J Mater 15:767–778
9. Gadouini H, Nadot Y, Rebours C (2008) Influence of mean stress on the multiaxial fatigue behaviour of defective materials. Int J Fatigue 30(9):1623–1633

10. Nadot Y (1997) Influence des défauts de fonderie sur la résistance à la fatigue d'une fonte GS, These, ENSMA, Poitiers
11. Chaboche JL (1981) Continuum damage mechanics-a tool to describe phenomena before crack initiation. Nucl Eng Des 64:233–247

Effects of WAAM Process Parameters on Metallurgical and Mechanical Properties of Ti-6Al-4V Deposits

A. Ayed[1(✉)], A. Valencia[1], G. Bras[2], H. Bernard[2], P. Michaud[2], Y. Balcaen[1], and J. Alexis[1]

[1] Laboratoire Génie de Production, INP ENIT, Tarbes, France
{achraf.ayed, amelie.valencia, yannick.balcaen, joel.alexis}@enit.fr

[2] University of Bordeaux, ESTIA Institute of Technology, 64210 Bidart, France
{g.bras, h.bernard, p.michaud}@estia.fr

Abstract. Additive manufacturing is a revolution for many sectors of the industry. These new manufacturing processes allow a substantial saving of raw material while optimizing the geometry, and at the same time, reducing development costs, by reducing the time between the concept and deployment phases of a product. A 3D printing device using a Cold Metal Transfer (CMT) arc welding station to melt a metallic filler wire is developed to build titanium parts by optimizing the process parameters to control metallurgical and mechanical properties of parts. In this study, two parameters (wire feed speed and movement speed of the robot) have been studied. Their impact on the metallurgical, dimensional process stability and mechanical properties of materials have been analyzed. Microstructure and mechanical properties vary depending on the energy expended during manufacture. This energy remains constant or decreases respectively when the wire feed speed or the robot head travel speed increases. Indeed, the electric generator adapts its power according to the speed of the wire. Regardless of the energy parameters, the movement speed of the robot seems to influence the wetting angles, the depth of melted zones, remelted and heat affected zones and the metallurgy with a refinement of the substructure of the deposits.

Keywords: Ti-6Al-4V alloy · Wire and arc additive manufacturing · Cold metal transfer · Microstructure

1 Introduction

Additive Manufacturing (AM) is a promising way to produce near-net shape metallic parts with complex geometries. It offers many advantages compared to machining processes, as reductions in the manufacturing lead-time and cost due to low material waste [6].

The Wire and Arc Additive Manufacturing (WAAM) is a Direct Energy Deposition (DED) additive manufacturing process derived from welding that uses a wire as filler metal and an electrical arc as heat source to produce wire melting [13]. It is a promising

F. Chaari et al. (eds.), *Advances in Materials, Mechanics and Manufacturing*,
Lecture Notes in Mechanical Engineering,
https://doi.org/10.1007/978-3-030-24247-3_4

process, especially because of its high deposition rate [3, 9], high energy efficiency [4], low-cost of raw materials, low material losses, and its capability to manufacture large parts [12]. Among the various arc processes that can be used for wire melting, the Cold Metal Transfer (CMT) process seems to be one of the more suited for WAAM, thanks to its controlled current waveform and filler wire feeding, that allow to obtain regular deposited weld bead and a mechanical control of the detachment of the molten drops ensuring the deposition of weld seams without or with very few projections [7].

Currently, few studies have been done on single-deposits Ti-6Al-4V produced by CMT. Most of the studies on CMT Ti-6Al-4V have been done on walls, and the impact of layer superposition on microstructure and mechanical properties has been studied. Ding et al. [3] show that the height of the seams increases and their width decreases with the translation speed of the laser. Zhang et al. observes a decrease in the penetration depth of the bead and of the wetting angles as a function of the speed of the robot by CMT on magnesium alloys. The present work is a contribution to the knowledge of the effects of these parameters (movement speed of the robot and wire feed speed) on the metallurgical and mechanical properties of material added on parts made by the CMT process to bring new functions.

2 Experimental Procedure

2.1 Description of Process

The cell is composed of a 6 axis robot (KUKA KR100-2 HA 2000) with its controller (KUKA KR C2-05 AK9) and a welding station (FRONIUS TPS 3200 CMT Remote) and a wire feeder (FRONIUS VR 7000-CMT 4R/G/W/F++). We choose a Ti-6AL-4V wire from the supplier Technalloy, its diameter is 1.2 mm. The welding process is inside an argon inserted chamber to ensure the protection of the Titanium weld bead against oxidation. The chamber design is based on the fact that Argon is heavier than the air: by pushing argon continuously in the box bottom, we progressively expel the air definitively. The chosen welding process is a derivate MIG/MAG process developed by FRONIUS and it's called CMT [5]. We choose the synergic law 1007 from the data-base R0981. The synergic law combines the wire feed speed and the delivered energy. Our experimental approach is to separate the robot speed and the wire feed speed. For that purpose, we print 10 seams from 0.5 to 1.3 cm s^{-1} for movement speed of the robot and 10 seams from 6 to 10.5 m min^{-1} for the wire feed speed. Current and energy measurements are measured and supplied directly by the FRONIUS machine.

2.2 Characterization Techniques

Dimensional analyzes of seams were measured with 3D ATOS scanner and analysed with CATIA V5 and ABviewer 12 software. All the seams were mechanically sectioned along the longitudinal and transverse directions from the central part (Fig. 1). The specimens were prepared in epoxy mount, ground with 600, 1200, 2500 grit SiC paper under water, polished with OPS solution with hydrogen peroxide to a mirror finish. They were then etched with the Kroll attack reagent composed of 3 mL of 40%

hydrofluoric acid, 2 mL of 68% nitric acid and 95 mL of distilled water. Matter health and microstructure were observed by optical (Leica wild M420 binocular and Olympus PMG3 microscope) and scanning electron microscopy (SEM JEOL JSM-7000F) on polished and etched specimens. EBSD analyzes (Oxford Nordlys II F+ camera) were done to determine the fine microstructure of each zone of the seam. Finally, longitudinal and transverse Vickers hardness filiations were made using a ZWICK ZHU 2.5 durometer with a load of 5 kg.

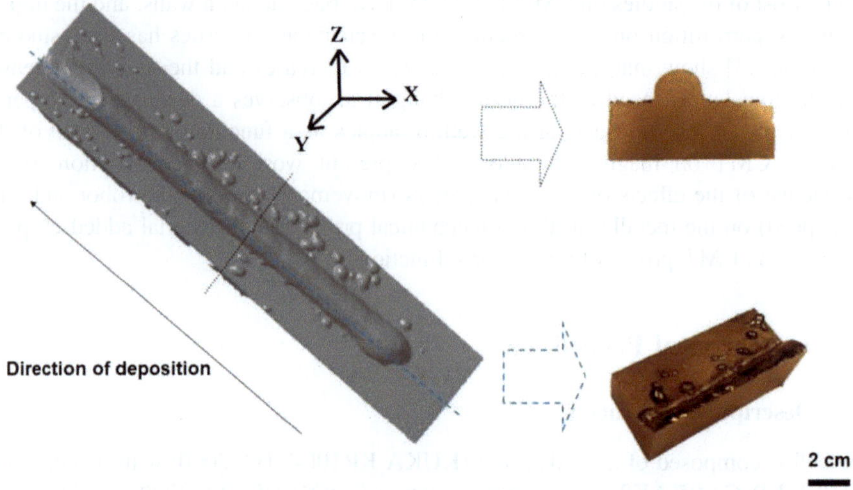

Fig. 1 Transverse (YZ) and longitudinal (XZ) sampling sections

3 Results and Discussion

3.1 Deposits Geometry (Shape and Size)

The morphology of the seams was first observed by optical microscopy before cutting (Fig. 2). The geometry of the seams is regular. They do not exhibit coloration induced by the phenomenon of oxidation. Projections are observable on the edges of some seams. In both cases, sampling according to the YZ and XZ plans were carried out as shown above in order to have as much information as possible on the evolution and the stability of the process.

The height of the seams increases with the wire feed speed and decreases with the speed of translation of the robot. It is the same for the width of the seams (Fig. 3). In conclusion, the amount of material deposited increases with the wire feed speed and decreases with the movement speed of the robot.

It was noticed a singular behavior of the deposit with a wire feed rate of 8.5 m/min compared to all the deposits.

This is due to an unexplained increase in projections when printing this deposit, which merged with the deposit increasing its width locally at the expense of its height.

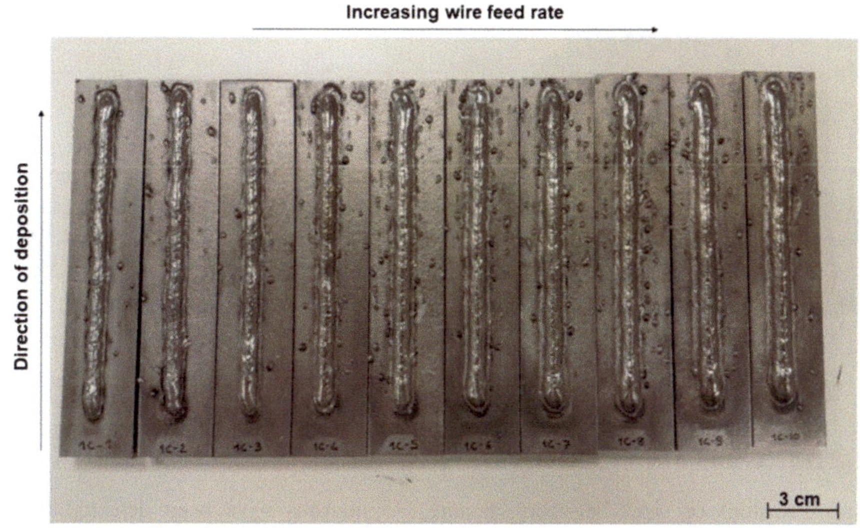

Fig. 2 Ti-6Al-4V deposits manufactured by varying wire feed speed speed

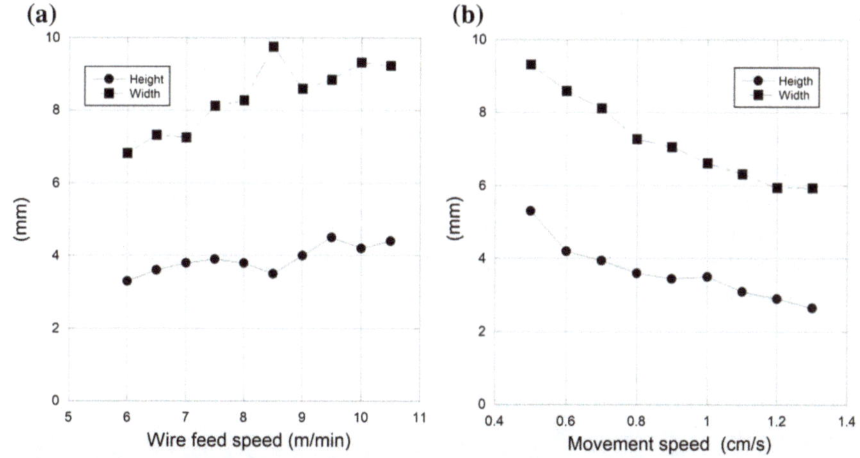

Fig. 3 Evolution of the depth of the main zones according to the wire feed speed (**a**) and robot's head speed (**b**)

The variation of these two parameters does not affect the stability of the process, where a similar regularity of the deposits and an almost constant number of projections have been noticed above 7.5 m min^{-1} of wire feed rate.

The wetting angles were also measured. It has been found that the wire feed speed does not affect these parameter, however the increase in the movement speed tends to increase the wetting angle (Fig. 4). Ding et al., conclude also that increasing the speed of the robotic head causes an increase in the wetting angle for Ti-6Al-4V alloy [3].

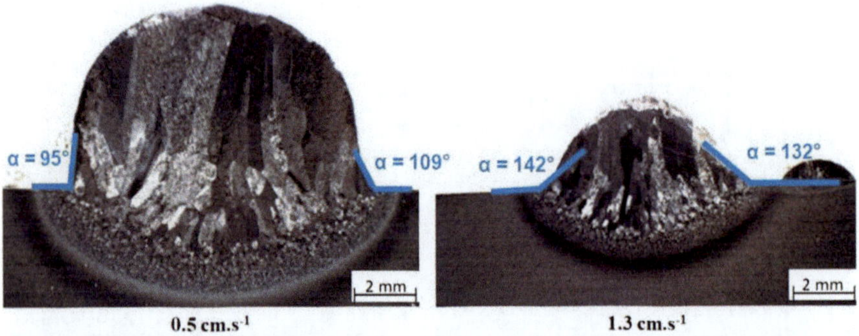

Fig. 4 Evolution of the wetting angle according to the movement speed of the robot

3.2 Defects and Microstructure

Some porosities of sizes between 16 and 375 microns have been detected at the deposit/remelted zone interface and in the remelted zone. These porosities are generally spherical, distributed in filiation or punctually. The characterization according to the transverse plane YZ showed a small constant porosity rate all along the process deposition.

The macrostructural study shows prior-β grains oriented in the direction of the thermal gradient. This type of microstructure was observed for Ti-6Al-4V by Martina [10]. The width of these β grains is stable as a function of the wire feed speed with average widths between 0.4 and 0.6 mm for all the deposits and an orientation relative to the vertical between 77° and 81°. On the other hand, a decrease in the width of the ex-grains β as a function of the movement speed of the robot has been observed, passing from an average width of 0.56 mm for the lowest speed to 0.34 mm for the highest speed (Fig. 5), this decrease reflects faster cooling rate. The orientation of these prior-β grains is between 77° and 84°.

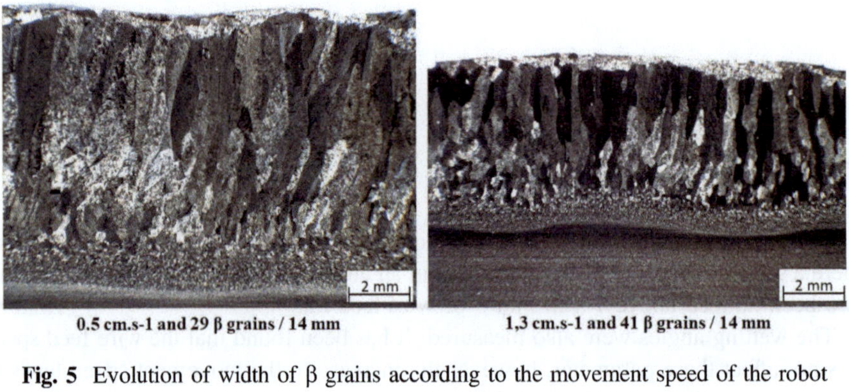

Fig. 5 Evolution of width of β grains according to the movement speed of the robot

The seams manufactured by CMT are composed of three different regions as shown in the Fig. 6. These regions are (from top to bottom): the melted zone, the remelted zone and the heat affected zone, the lower is the base metal.

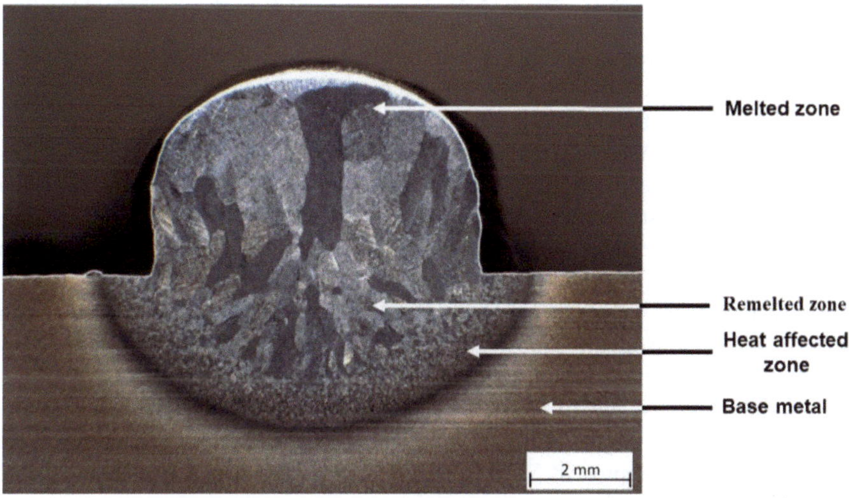

Fig. 6 Micrography of cross section of Ti-6Al-4V deposit

It has been noticed an evolution of the dimensions of these zones according to the parameters (Fig. 7). Indeed, in the case of the variation of the wire feed speed, the height of the deposits increases and confirms the dimensional analyzes made before. The deposit with the wire speed of 8.5 m min^{-1} has a very large width because of the many projections merged with the deposit. On the other hand, the depth of remelted

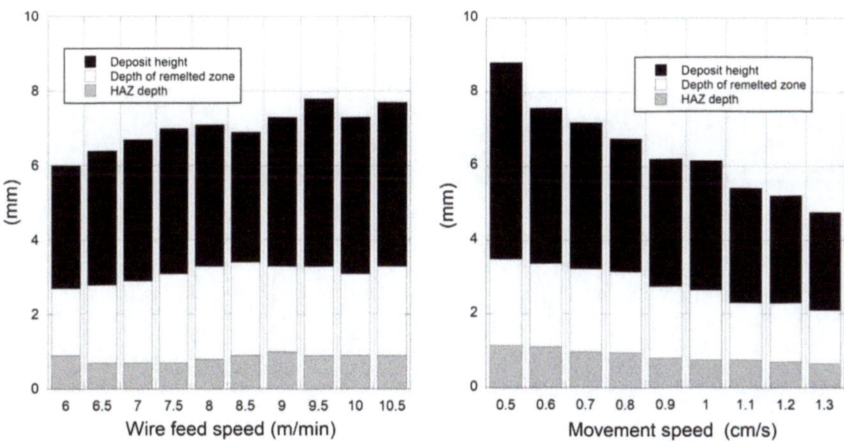

Fig. 7 Evolution of the depth of the main zones according to the wire feed speed and movement speed of robot

zone and heat affected zones (HAZ) are quite similar varying from 1.5 to 2.5 mm for the remelted zone and 0.7 to 0.9 mm for the heat affected zone (HAZ). The remelted zone and HAZ depth, however, varies more according to the movement speed from 5.2 to 2.7 mm for the remelted zone and from 1.15 to 0.6 mm for the HAZ for the fastest robot speed. These observations were observed by Zhang [14] for CMT applied to a magnesium alloy.

The microstructural study shows a microstructure of Widmanstätten for all the deposits studied, which indicates a high cooling rate during the beta/alpha transformation, of the order of 8000 °C/min [8]. This type of microstructure has been observed by Antonysamy and Wang [1, 11]. So, the size of the lamellae α does not vary as a function of the wire feed speed. A slight variation in the size of the α-lamellae, which are finer for the most important movement speeds of robot as shown below which confirms the existence of relations between movement speed and Ti-6Al-4V cooling rates (Figs. 8 and 9).

It has also been noticed a gradient of α lamellae width which are increasingly thinner at the bottom of the deposits compared to their upper parts as shown in the Electron Back Scatter Diffraction (EBSD) analysis below.

This difference is due to the Ti-6Al-4V support plate on which the seam is deposited, and which plays the role of heat sink, enabling the conduction of the heat generated by the CMT process in the base zone of the deposits.

3.3 Mechanical Properties

The hardness of deposits is presented in the figure below (Fig. 10).

The wire feed speed does not affect the hardness of the as-built deposits where-as increasing the speed of the robot head induces a light increase of hardness of deposits up to a robot speed of 0.9 cm s^{-1}, beyond the hardness remains constant. However, a threshold is reached around 350–360 HV. These results are similar to those found by Brandl [2]. The hardness of the HAZ is always greater than the hardness of melted and remelted zones.

Fig. 8 Microstructure of deposits 1C-11 (left) and 1C-19 (right)

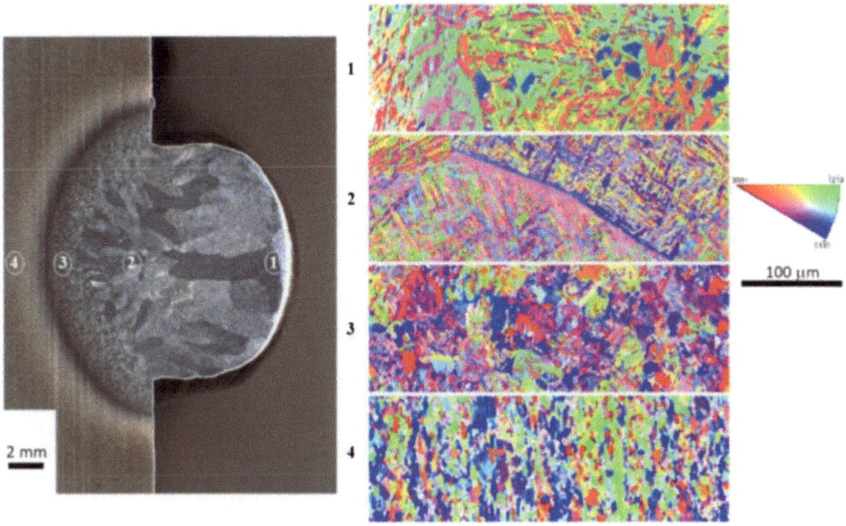

Fig. 9 Inverse Pole Figure of cross-section obtained by EBSD analysis of a Ti-6Al-4V CMT deposit

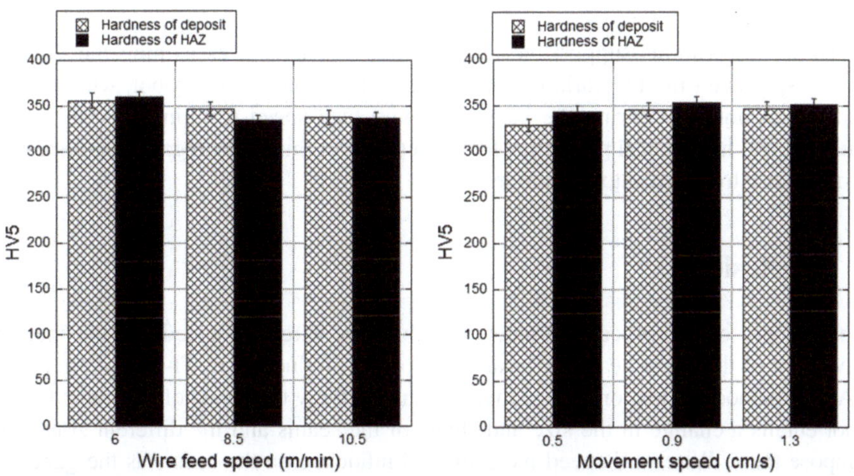

Fig. 10 Mean hardness of the Ti-6Al-4V deposits according to the wire feed speed and movement speed of robot

In order to have a better understanding of the phenomena observed, a study of the energy parameters is carried out. The Fronius generator automatically adjusts the energy density delivered during wire feed rate variation. On the contrary, it is not adjusted during the evolution of the head speed; the energy per volume of deposited material decreases according to the movement speed as shown in the Fig. 11.

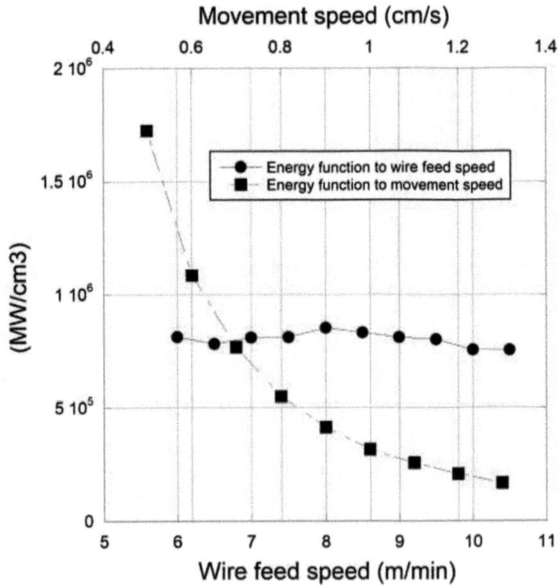

Fig. 11 Energy density delivered per volume of deposited material during the CMT process

These observations support the results found previously, explaining the different cooling speeds during the variation of the movement speed of the robot, which on the contrary do not vary during the increasing wire feed speed resulting in similar metallurgical and mechanical properties for all deposits of this study regardless of the flow of material introduced into the molten bath.

4 Conclusion

The influence of two process parameters such as the wire feed speed and the speed of movement of the robot head was evaluated for the production of Ti-6Al-4V seams by WAAM method. Oxidation-free seams were made. Only the speed of movement of the robot entails a change in the size and shape of the seams and the different zones that compose them. Wire feed speed has a limited influence on the seams as the generator compensates by varying the power density. The microstructural evolutions and the properties of the seams are thus linked to the energy variations per unit of material deposited during the process.

Acknowledgements. This work was financially supported by the FUI program (FUI Addimafil). The authors thank Jade Pécune for her help to the metallographic preparation.

References

1. Antonysamy AA (2012) Microstructure, texture and mechanical property evolution during additive manufacturing of TI-6Al-4V alloy for aerospace applications. The university of Manchester
2. Brandl E, Baufeld B, Leyens C, Gault R (2010), Additive manufactured Ti-6Al-4V using welding wire: comparison of laser and arc beam deposition and evaluation with respect to aerospace material specifications. ScienceDirect. https://doi.org/10.1016/j.phpro.2010.08.087
3. Ding D, Pan Z, Cuiuri D, Li H (2015) Wire-feed additive manufacturing of metal components: technologies, developments and future interest. Int J Adv Manuf technol 81:465–81 (2015). https://doi.org/10.1007/s00170-015-7077-3
4. Dupont JN, Marder AR (1995) Thermal efficiency of arc welding processes. Weld J-Weld Res Suppl 74:406s
5. Fronius International GMBH (2018)CMT—cold metal transfer: the cold welding process for premium quality. http://www.fronius.com/en/welding-technology/our-expertise/welding-processes/cmt. Accessed 11 Sept 2018
6. Frazier WE (2014) Metal additive manufacturing: a review. J mater Eng Perform 23:1917–1928. https://doi.org/10.1007/s11665-014-0958-z
7. Gomez Ortega A (2018) Prototypage rapide de pièces en alliage d'aluminium: étude du dépôt de matière et d'énergie lors de la fusion à l'arc d'un fil par le procédé MIG-CMT. Université de Montpellier
8. Lütjering G (1998) Influence of processing on microstructure and mechanical properties of (α+β) titanium alloys. Mater Sci Eng A243: 32–45
9. Martina F, Mehnen J, Williams SW, Colegrove P, Wang F (2012) Investigation of the benefits of plasma deposition for the additive layer manufacturing of the Ti-6Al-4V. J Mater Process Technol 212(6):137–1386
10. Martina F, Colegrove PA, Williams ST, Meyer J (2012) Microstructure of Interpass rolled wire+arc additive manufacturing Ti-6Al-4V Components. Metall Mater Trans A. https://doi.org/10.1007/s11661-015-3172-1
11. Wang F, Williams S, Colegrove P, Antonysamy AA (2013) Microstructure and mechanical properties of wire and arc additive manufactured Ti-6Al-4V. Metall Mater Trans A 44A:977. https://doi.org/10.1007/s11661-012-1444-6
12. Williams S, Martina F (2015) Wire+Arc additive manufacturing vs. Traditional machining from solid: a cost comparison. Technical report. Welding Engineering and Laser Processing Centre, Cranfiels University
13. Williams SW, Martina F, Addison AC, Ding J, Pardal G, Colegrove P (2016) Wire+Arc additive manufacturing. Mater Sci Technol 32:641–647. https://doi.org/10.1179/1743284715y.0000000073
14. Zhang H, Hu S, Wang Z, Liang Y (2015) The effect of welding speed on microstructures of cold metal transfer deposited AZ31 magnesium alloy clad. Mater Des 86:894–901

Structural Characterization of Phosphate-Based Geopolymer

M. Zribi(✉), B. Samet, and S. Baklouti

Laboratoire de Chimie Industrielle, Ecole Nationale d'Ingénieurs de Sfax, Université de Sfax, BP.W3038 Sfax, Tunisia
{zribi.mar, baklouti.samir}@gmail.com, sametbasma@yahoo.fr

Abstract. Geopolymers obtained following acidic activation represent a promising future material with their different advantages, namely high mechanical strength, excellent dielectric properties and important chemical stability. In the present research, Phosphate-based geopolymer was prepared from metakaolin, as the aluminosilicate precursor and commercial phosphoric acid H_3PO_4, as the activator precursor. These reagents were mixed with (Al/P) molar ratio and solid/liquid (S/L) volume ratio equal to 1. To investigate the obtained material structure, different techniques were used as Fourier Transform Infrared spectroscopy (FTIR), magic angle spinning nuclear magnetic resonance (MAS-NMR) spectra of ^{29}Si, ^{27}Al and ^{31}P, X-ray diffraction powder (XRD) and scanning electron microscopy (SEM). As well, the same techniques were used to characterize the used aluminosilicate precursor, which is metakaolin. As a result, the formation of a first aluminum phosphate geopolymeric network dispersed in a second one based on (Si–O–T) units was proven, with T=Si, Al and P. The XRD analyze shows the formation of an amorphous material. In fact, any crystalline phases were observed. Accordingly, the Phosphate-based geopolymer was considered as a composite material composed from two amorphous networks: the aluminum phosphate geopolymeric network which plays the role of reinforcement and the silicate network which represents the matrix.

Keywords: Phosphate-based geopolymer · Structure · Composite · Metakaolin · Characterization

1 Introduction

Phosphate-based geopolymers are new eco-materials obtained by the reaction of aluminosilicate precursor with phosphoric acid solution. Recently, this family of material becomes more and more attractive. This importance is explained by the different advantages of this material namely high mechanical strength [1–3], excellent dielectric properties [4], and important chemical stability [3].

These properties are very dependent on the material's structure. Investigating the literature, this structure is not yet well defined. The first structure was proposed by [5]. It was based on the presence of (–Si–O–P–O–Al–) units. This structure was confirmed by many authors [6, 7]. Recently, Louati et al. [1] propose another structure. The authors proposed the formation of a geopolymeric composite composed of a crystalline

F. Chaari et al. (eds.), *Advances in Materials, Mechanics and Manufacturing*, Lecture Notes in Mechanical Engineering,
https://doi.org/10.1007/978-3-030-24247-3_5

phase of $AlPO_4$ dispersed in an amorphous structure of (–Si–O–P–) geopolymeric units. In the same time, many authors continue to adopt the formation of (–Si–O–P–O–Al–) units like He et al. [3] and Wang et al. [8].

This research aims to investigate the structure of Phosphate-based geopolymer following different techniques as Fourier Transform Infrared spectroscopy (FTIR), magic angle spinning nuclear magnetic resonance (MAS-NMR) spectra of ^{29}Si, ^{27}Al and ^{31}P, X-ray diffraction powder (XRD) and scanning electron microscopy (SEM). With each technique, the characterization of the used aluminosilicate precursor, which is metakaolin, is also mentioned.

2 Materials and Experimental Methods

In this work, the Phosphate-based geopolymer was obtained from the metakaolin activation. The used metakaolin was provided by Imerys France. In addition, commercial phosphoric acid (H_3PO_4 85% by mass) provided by Scharlau-chémie (SA) society was used as activator. To synthesize the studied material, metakaolin, phosphoric acid and distilled water were mixed with (Al/P) molar ratio and solid/liquid (S/L) volume ratio equal to 1. The obtained mixture was kept at a temperature of 60 °C in a plastic mold. After 24 h of curing, the material becomes consolidated and different structural characterizations were carried: Fourier- transform infrared spectroscopy (FTIR) was carried by a Perkin Elmer spectrum BX spectrophotometer apparatus in transmittance mode and wave number range between 4000 and 450 cm^{-1}. In addition, the magic angle spinning nuclear magnetic resonance (MAS-NMR) spectra of ^{29}Si, ^{27}Al and ^{31}P were performed out by using BRUKER 300 WB (7T), to investigate the chemical environments. ^{31}P, ^{29}Si and ^{27}Al shifts are given in ppm referenced to H_3PO_4, TMS and aluminum chloride, respectively. As well, X-ray diffraction powder (XRD) with a BRUKER-AXS-D8-Advance powder diffractometer using Cu Kα radiations (λkα = 1.5418 Å) was carried out. The analytical range was between 5° and 70° (2θ) at a rate of 1°/min. Finally, the scanning electron microscopy (SEM) was carried out using Jeol JSM-5400 microscope on sample, to observe the new microstructure.

3 Results and Discussion

3.1 Infrared Spectroscopy (FTIR)

The infrared spectra of both metakaolin and Phosphate-based geopolymer are illustrated in Fig. 1. The metakaolin spectrum shows the presence of a first intense band located at 1076 cm^{-1}. It is attributed to Si–O stretching vibration [9]. Another band located at 800 cm^{-1} is observed. This band is assigned to (Si–OH) bending vibration [10]. As well, the spectrum shows a band at 448 cm^{-1}, which can be attributed to bending vibrations of (Al–OH) linkages. After geopolymerisation, Phosphate-based geopolymer spectrum presents some changes, due to the new structure: the first remarkable change is the disappearance of the bands located at 800 and 448 cm^{-1}, respectively. This disappearance proves the condensation between monomeric silanol (Si–OH) and aluminol (Al–OH) units to form the new polymeric structure. Added to that, the band located at 1076 cm^{-1} shifted from 1076 to 1064 cm^{-1}. This shift can be explained by the transformation in the Si layer, to form the new geopolymeric network.

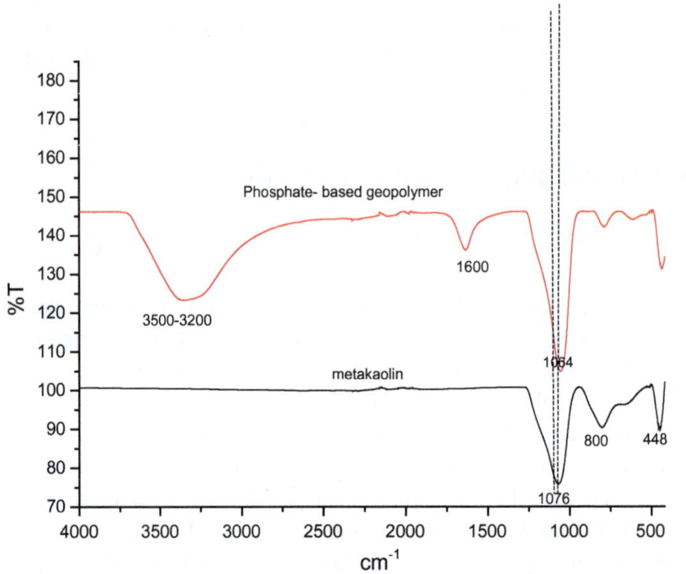

Fig. 1 Infrared spectra of both metakaolin and phosphate-based geopolymer

3.2 Nuclear Magnetic Resonance (NMR)

The solid-state NMR analysis is one of the most used techniques in the determination of geopolymeric structures. The ^{29}Si spectra of both metakaolin and Phosphate-based geopolymer are illustrated in Fig. 2a. The ^{29}Si spectrum of metakaolin shows a single large signal around $\delta = 92.05$ ppm assigned to the Q3 (3Si, 1OH) environment [7]. This signal characterizes, usually, the amorphous metakaolin structure. After geopolymerisition, a slight shift in this signal is observed from $\delta = -92.05$ ppm to $\delta = -116.54$ ppm. This signal with this new resonance correlated with Q4 environment [7]. The observed shift can be explained by the transformation of the Si environment from (3Si, 1OH) to (4Si) [11]. We talk about the formation of the first 3D geopolymeric network Si–O–Si in the studied structure. A second signal located around $\delta = -101.35$ ppm is assigned to Q3 environments with (3Si, 1Al). This signal shows the persistence of some dealuminated aluminum attached on the starting Si layer. Another narrow shoulder is observed at $\delta = -92.05$ ppm which correlated with the Q3 environment in the residual metakaolin. The weakness of this peak improves the consumption of the starting metakaolin to form the new geopolymeric structure.

The ^{27}Al spectrum of metakaolin is illustrated in Fig. 2b. This spectrum shows the presence of a large and antisymmetric peak. This peak characterized an amorphous

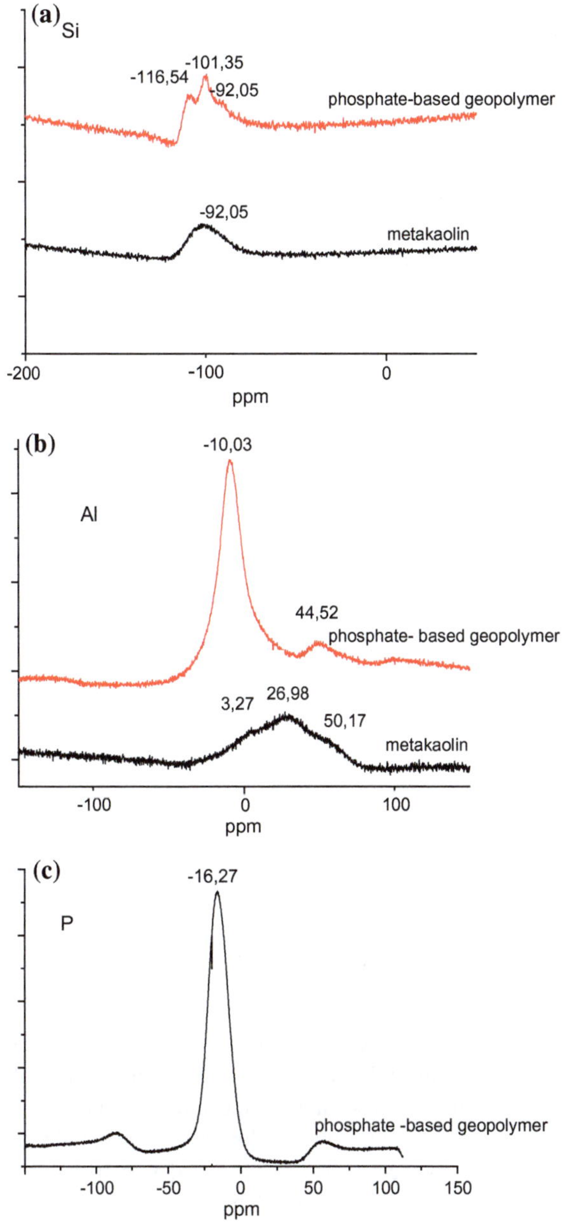

Fig. 2 ^{29}Si (**a**), ^{27}Al (**b**) and ^{31}P (**c**) NMR spectra

metakaolin with the presence of three aluminum environments: the aluminum hexa-coordinated at 3.27 ppm, the aluminum penta-coordinated at 26.98 ppm and the aluminum tetra coordinated at 50.17 ppm [11]. After geopolymerisation, two resonances are observed at $\delta = 44$ ppm and $\delta = -10$ ppm, respectively. The first resonance located at 44 ppm is assigned to tetrahedral coordinated Al environment. The second resonance located at −10 ppm coincides with the Al hexa coordinated [12]. These resonances suggest the presence of another geopolymeric network based on P–O–Al–O– units [13].

The analyze of ^{31}P spectra of geopolymer, illustrated in Fig. 2c, confirms again the presence of the second discussed geopolymeric network. This spectrum shows the presence of only one resonance at around −16 ppm. This resonance correlated with the same order as for the polymerized phosphorus species environment [14]. It confirms the presence of the observed aluminum phosphate network, already discoursed with ^{27}Al spectra.

3.3 X-Ray Diffraction

The XRD analysis of metakaolin, illustrated in Fig. 3, shows the presence of a diffuse halo between 18° and 30° characteristic of an amorphous material obtained by calcination of kaolinite. The absence of any crystalline phase in the used metakaolin is proved by the absence of any peaks in the corresponding spectrum.

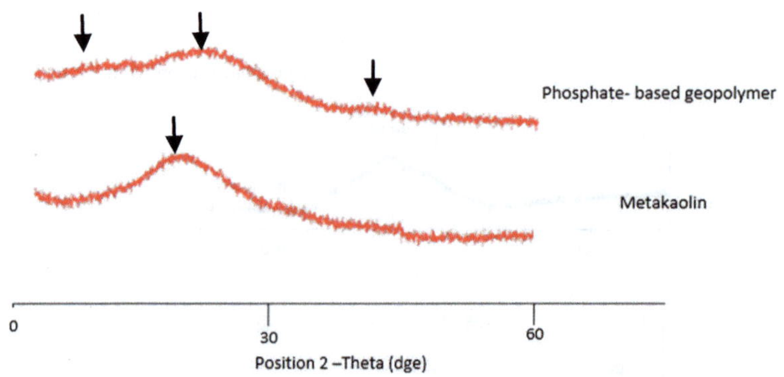

Fig. 3 X-ray diffraction spectra of metakaolin and phosphate-based geopolymer

After geopolymerisation, the XRD analysis of the obtained material shows the appearance of two new diffuse halos, compared to the metakaolin spectrum. As illustrated in Fig. 3, the first diffuse halo is located between 5° and 15°. It can be attributed to the presence of an amorphous aluminuim phosphate phases [15]. Moreover, another diffuse halo located at 45° is observed which justifies the presence of another amorphous geopolymeric network. Therefore, the obtained structure corresponds to a composite material composed from two amorphous networks.

3.4 Scanning Electron Microscopy

The SEM examinations of Phosphate-based geopolymer aged 1 day, is illustrated in Fig. 4. The SEM image shows the obtaining of a consolidated material. Besides, the presence of the two distinct networks is observed, as mentioned in the figure:

Aluminum phosphate phases are dispersed in another geopolymeric network to play the role of reinforcement [9].

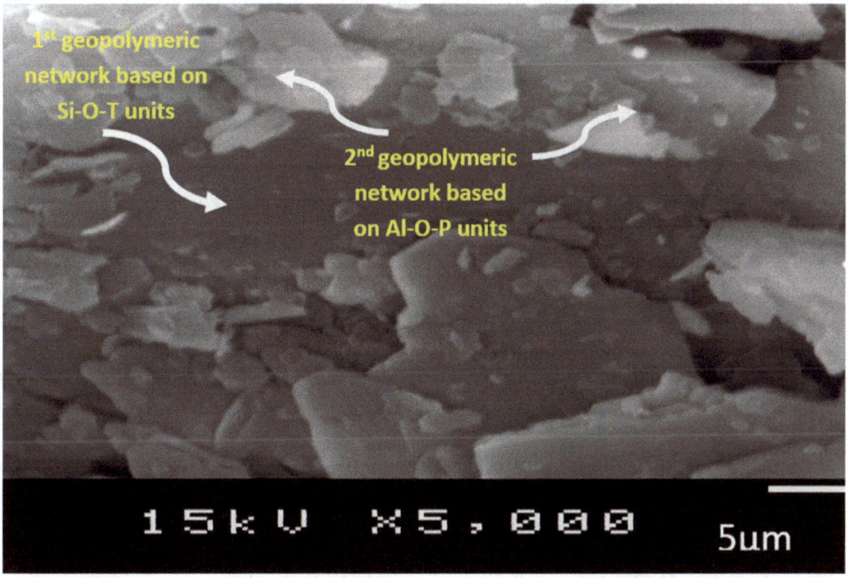

Fig. 4 SEM image of phosphate-based geopolymer aged 1 day of curing

4 Conclusion

The combination of the Fourier Transform Infrared spectroscopy (FTIR), magic angle spinning nuclear magnetic resonance (MAS-NMR) spectra of ^{29}Si, ^{27}Al and ^{31}P, X-ray diffraction powder (DRX) and scanning electron microscopy (SEM) provided the formation of an amorphous composite material. It is composed of a first aluminum phosphate geopolymeric network dispersed in a second one based on (Si–O–T) units, with T=Si, Al and P. The obtained material is totally amorphous and it is free from any crystalline phases.

References

1. Louati S, Baklouti S, Samet B (2016) Acid based geopolymerization kinetics: effect of clay particle size. Appl Clay Sci 132–133:571–578. https://doi.org/10.1016/j.clay.2016.08.007404
2. Tchakouté HK, Rüscher CH (2017) Mechanical and microstructural properties of metakaolin-based geopolymer cements from sodium waterglass and phosphoric acid solution as hardeners: a comparative study. Appl Clay Sci 140:81–87
3. He Y, Liu L, He L, Cui X (2016) Characterization of chemosynthetic H_3PO_4–Al_2O_3–$2SiO_2$ geopolymers. Ceram Int 42:10908–10912

4. Douiri H, Louati S, Baklouti S, Arous M, Fakhfakh Z (2014) Structural, thermal and dielectric properties of phosphoric acid-based geopolymers with different amounts of H_3PO_4. Mater Lett 116:9–12
5. Cao D, Su D, Lu B, Yang Y (2005) Synthesis and structure characterization of geopolymeric material based on metakaolinite and phosphoric acid. Guisuanyan Xuebao (J Chin Ceram Soc)
6. Le-Ping L, Xue-Min C, Shu-Heng Q, Jun-Li Y, Lin Z (2010) Applied clay science preparation of phosphoric acid-based porous geopolymers. Appl Clay Sci 50:600–603. https://doi.org/10.1016/j.clay.2010.10.004
7. Perera DS, Hanna JV, Davis J, Blackford MG, Latella BA, Sasaki Y, Vance ER (2008) Relative strengths of phosphoric acid-reacted and alkali-reacted metakaolin materials. J Mater Sci 43:6562–6566
8. Wang YS, Dai JG, Ding Z, Xu WT (2017) Phosphate-based geopolymer: formation mechanism and thermal stability. Mater Lett 190:209–212. https://doi.org/10.1016/j.matlet.2017.01.022
9. Tchakouté HK, Rüscher CH, Kamseu E, Andreola F, Leonelli C (2017) Influence of the molar concentration of phosphoric acid solution on the properties of metakaolin-phosphate-based geopolymer cements. Appl Clay Sci 147:184–194. https://doi.org/10.1016/j.clay.2017.07.036
10. Sadangi JK, Muduli SD, Nayak BD, Mishra BK (2013) Effect of phosphate ions on preparation of fly ash based geopolymer. IOSR J Appl Chem 4:20–26
11. Davidovits J (2008) Geopolymer chemistry and applications, 2nd Ed. Saint-Quentin, Fr. Inst. Geopolymer
12. Zhang L, Eckert H (2004) Sol–gel synthesis of Al_2O_3–P_2O_5 glasses: mechanistic studies by solution and solid state NMR. J Mater Chem 14:1605–1615
13. Davidovits J (2015) Geopolymer chemistry and applications. 2nd Ed. Saint-Quentin, Fr. Inst. Geopolymer
14. Mudrakovskii IL, Shmachkova VP, Kotsarenko NS, Mastikhin VM (1986) 31P NMR study of I-IV group polycrystalline phosphates. J Phys Chem Solids 47:335–339
15. Gaspar SG (2013) Ph.D. Thesis. Mise en forme par extrusion de supports de catalyseurs à base d'alumine et à microstructure multi-échelles: effet de la composition granulaire et du liant sur les propriétés des matériaux. Institut National des Sciences Appliquées de Lyon

Influence of Tool-Workpiece Characteristics on Cutting Tool Dynamic Behaviour and Its Position During Straight Turning

Romdhane Othmani[1(✉)] and Mourad Saidi[2]

[1] UGPM2, Mechanical Engineering and Materials Production Unit, Sfax, Tunisia
othmani_romdhane@yahoo.fr

[2] UR-MSSDT, Mechanics of Solids, Structures and Technological Developments, Tunis, Tunisia
mouradsaidi840@gmail.com

Abstract. Production in industry is limited by the vibration phenomenon, but it is also at the root of the problems of surface quality and degradation of the cutting tool and the machine. The vibration has been widely studied and modeled. Most models aim to model the machining system using a model with a single degree of freedom or two degrees of freedom where the cut is stable depending on the dynamic properties of the part—tool—machine. The main objective of this article is to model numerically in three dimensional the dynamic behaviour of the machining system during turning operation while considering the vibration of the unit workpiece and cutting tool. The writing matrix of the differential equation of movement of the system was developed by neglecting their damping effect where the cutting tool and the workpiece are studied and characterized together by matrices mass and stiffness. The finite element method is used to obtain the global matrices mass and stiffness of the machining system by assembling the elementary matrices of the cutting tool and the workpiece by considering them as two elastic deformable beams. We have identified after the assembly that there is a matrix coupling. This method is also used to identify the eigens characteristics of the system using the Matlab software as a numerical simulation tool during a straight turning and to analyze the influence of the cutting tool position and its mechanical characteristics on the dynamic behaviour of the machining system. The dynamic model of the machine was obtained by a numerical modal analysis to determine the modal displacement of the machining system which is influenced by modifying the position of the cutting tool have a great influence on the behaviour dynamic and the modulus of elasticity which. Each machining system being machined has its own frequencies and modes, for frequencies the system is stable and others unstable. The paper concludes with numerical results for this three-dimensional numerical model show that in order to have a stable system, it is obligatory to avoid the eigens modes and frequencies of the machining system.

Keywords: Machining system · Global matrix · Finite element method · Eigen characteristics · Dynamic behaviour · Numerical simulation

F. Chaari et al. (eds.), *Advances in Materials, Mechanics and Manufacturing*,
Lecture Notes in Mechanical Engineering,
https://doi.org/10.1007/978-3-030-24247-3_6

1 Introduction

Manufacturing process presented many problems related to machining parameters [1]. In this context, during a turning operation there is vibration occurring between the tool and the workpiece which generates a poor surface condition. Xiao and al. point out that during vibration cutting the tool vibrations are damped, while during conventional cutting the vibrations are generated. It means that vibration cutting is more stable with respect to conventional one [2]. Experimental and simulation presented in demonstrate that vibration cutting helps to control tool vibration [3]. More models have been developed by researchers to explain the phenomenon of vibration, to predict the dynamic behaviour of the machining system according to the position of the tool [4] and to predict the optimal conditions for the development of a stable cut [5, 6]. We model from existing models a three-dimensional model of the dynamic behaviour of the machining system using the Finit Element Method (FEM) [7]. The resonance mode, maintaining the same amplitudes of tool vibrations is considered to be the best for vibration cutting [8]. The combined application of numerical analysis with accurate finite element model as well as different experimental methods during investigation of the vibration turning process allowed determining that the most favorable is the second flexural vibration mode of the tool [9]. All these dynamic models do not express correctly the dynamic behaviour of the machining system since they excluded the influence from the vibration of the workpiece. We model starting from the existing models of the dynamic behaviour of the system machining of use the FEM.

2 Numerical Model

To apply the Finite Element Method (FEM) to machining system, we consider several elements whose number depends on the precision that we want to achieve. We calculate for each of them the elementary matrices of mass K_e and stiffness M_e then we proceed to the calculation of global matrices; the global matrix M_{WT} is obtained by assembling the matrices of elementary masses and the stiffness matrix K_{WT} is obtained by assembling stiffness matrices elementary [6, 7].

The Finite Element Method (FEM) leads to obtaining a differential equation of global motion while neglecting the damping effect of the unit cutting tool and workpiece under the following matrix form:

$$\{M_{WT}\}\{\ddot{q}\} + \{K_{WT}\}\{q\} = \{0\} \tag{1}$$

With:

M_{WT} matrix global mass of the system machining workpiece-tool
K_{WT} matrix global stiffness of the system machining workpiece-tool
q vector moving point of contact between the workpiece and the tool
$\ddot{q}$ vector acceleration of the point of contact between the workpiece and the tool.

Figure 1 shows the 3D dynamic model of the machining system.

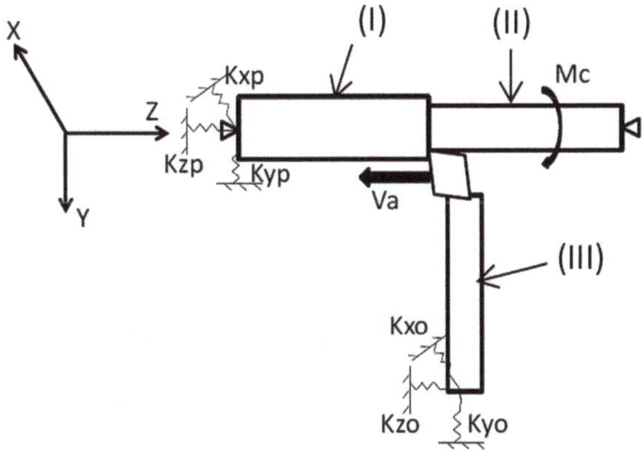

Fig. 1 3D dynamic model of the machining system

With;

The elements (I) and (II) correspond to the workpiece and element (III) corresponds to the cutting tool that are considered as stiffness beams K_{ij} and mass M_{ij}.

i = {x, y, z}: the three directions of space
j = {I, II, III}: element number (Fig. 2).

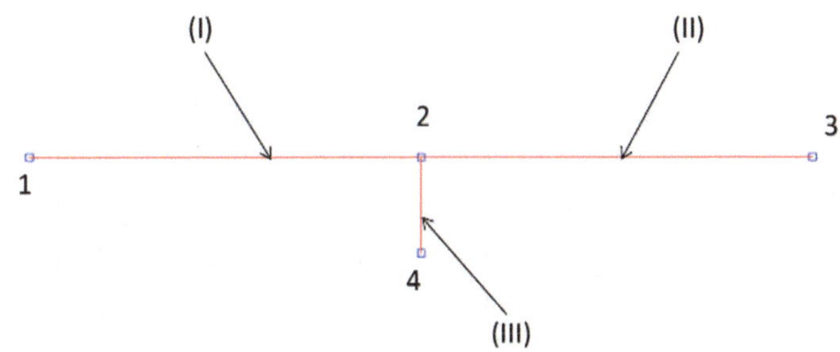

Fig. 2 Modelization of the machining system by finite element method

We modeled the workpiece by the Finite Element Method finished by a beam discretized in two elements (I and II), three knots and five degrees of freedom (X_k Y_k Z_k θ_{ijk} θ_{ijk}) for each node and the cutting tool is modeled by a single element (III) and two nodes.

With:

k = {right, left}: the two sections of each element.

2.1 Global Displacement Vector

After assembling the two vectors movements of the tool and the workpiece, based on compatibility conditions for travel level of the common node of the two elements, we get the global displacement vector of the block workpiece-tool q_{WT} of order (1 × 21) and are written in the following form:

$$q_{WT} = q_{1-2-3} \tag{2}$$

2.2 Matrix Global Stiffness

The stiffness matrix K_{WT} is deduced from the matrices K_W and K_T. The global stiffness matrix K_{WT} is order (21 × 21) and may consist of ten sub-matrices and it takes the form of next:

$$[K_{WT}] = \begin{bmatrix} [A_{(1-2-3)}] & [0] & [U] & [0] & [H] \\ [0] & [B_{g1}] & [E_1] & [0] & [0] \\ [U]^T & [E_1]^T & [C_{gd(1-2-3)}] & [E_2] & [F] \\ [0] & [0] & [E_2]^T & [D_{d2}] & [0] \\ [H]^T & [0] & [F]^T & [0] & [G_{g3}] \end{bmatrix} \tag{3}$$

2.3 Matrix Global Mass

The matrix global mass M_{WT} is deduced from matrices M_W and M_T. The global mass matrix M_{WT} is of order (21 × 21) and may consist of ten sub-matrices and takes the following form:

$$[M_{WT}] = \begin{bmatrix} [A''_{(1-2-3)}] & [0] & [U''] & [0] & [H''] \\ [0] & [B'_{g1}] & [E'_1] & [0] & [0] \\ [U'']^T & [E'_1]^T & [C''_{gd(1-2-3)}] & [E'_2] & [F''] \\ [0] & [0] & [E'_2]^T & [D'_{d2}] & [0] \\ [H'']^T & [0] & [F'']^T & [0] & [G''_{g3}] \end{bmatrix} \tag{4}$$

With;

The matrix A″ (1-2-3) is of order (4 × 4) and which corresponds to the traction-compression of the two elements of the workpiece and the bending of the tool. The matrix C″gd (1-2-3) corresponds to the flexion of the interface section and the compression-pull of the tool. The matrix B′g1 corresponds to the flexion of the element (1) of the room. D2 matrix corresponds to the bending of the element (2) of the tool.

The matrix G″g3 corresponds to the flexion of the element (3) of the tool. The following matrices: E′1, E′2, F″, H″, U″ correspond to the coupling matrices in the global mass matrix.

3 Numerical Results

3.1 Eigen Frequency Analysis

Table 1 represents the determination of the eigen frequencies of the machining system according to the position and the modulus of elasticity of the tool, we find that when the cutting tool is at the end of machining the eigen frequencies increase. Also, as the modulus of elasticity of the cutting tool increases, these frequencies increase.

Table 1 Determination of the eigen frequencies (F en 104 Hz) according to the modulus of elasticity and the position of the tool

Modulus of elasticity E_T (105 MPa)	Position of the tool	
	The beginning of machining	The end of machining
1	0.0398	0.2293
	0.1724	0.6980
	0.6855	0.8089
	0.9788	1.8842
	1.1335	1.3722
	1.4017	3.0410
2.1	0.0398	0.2356
	0.1787	0.9145
	0.9800	0.9929
	1.2950	1.2521
	1.3619	1.7342
	1.7033	3.0642

3.2 Modal Displacement Analysis

Figure 3 represents the determination of the modal displacement of machining system according to the position and the modulus of elasticity of the cutting tool. The analysis is dedicated only to displacements at the contact workpiece cutting tool. For the modulus of elasticity Eo = 1×10^5 MPa, the figure shows that at the beginning of machining, the 3rd mode and the 5th mode are the most dangerous and at the end of machining the 2nd and the 5th mode are the most dangerous. And for the other modulus of elasticity, the figure shows that at the beginning of machining, the 3rd mode and the 5th mode are the most dangerous and at the end of machining the 3rd and the 5th mode are the most dangerous.

<table>
<tr><th rowspan="2">Modulus of Elasticity E_T (105 MPa)</th><th colspan="2">Position of the cutting tool</th></tr>
<tr><th>The beginning of machining</th><th>The End of machining</th></tr>
<tr><td>1</td><td>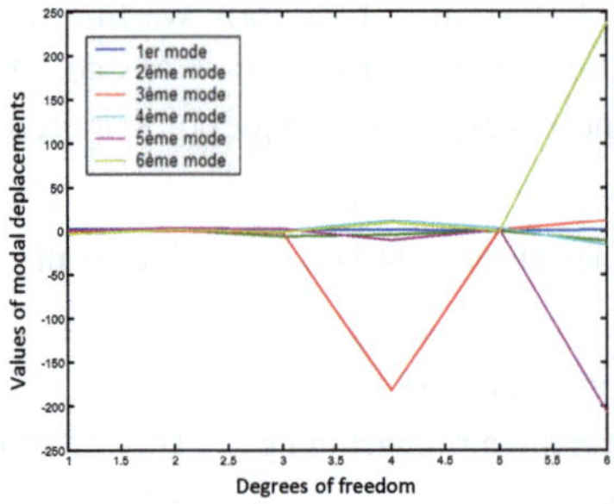
</td><td>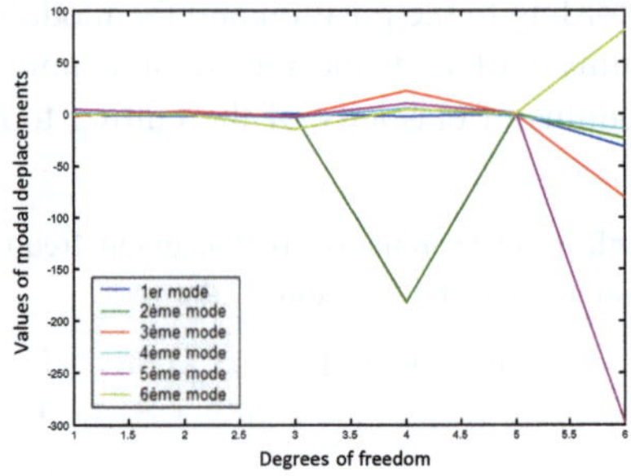
</td></tr>
<tr><td>2.1</td><td>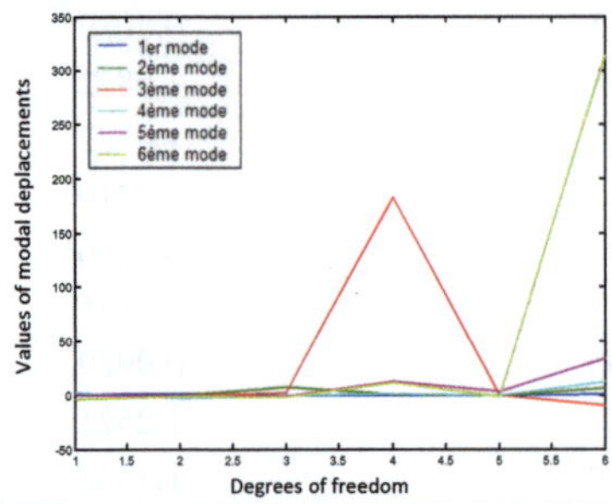
</td><td>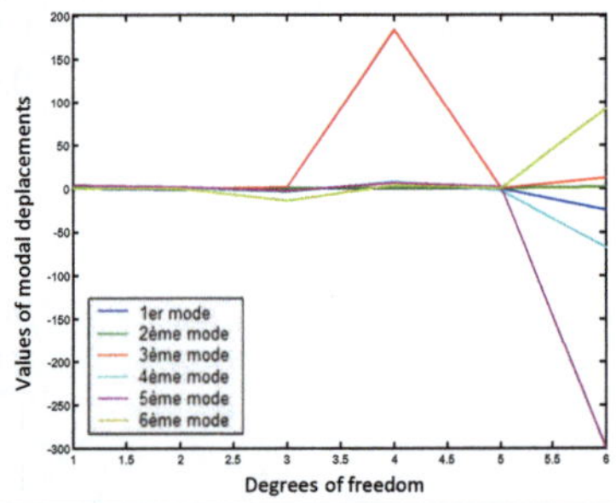
</td></tr>
</table>

Fig. 3 Determination of the modal displacement according to the modulus of elasticity and the position of the cutting tool

4 Conclusion

In this article, we have shown the variation of eigen frequencies and eigen modes, depending on the modulus of elasticity and the position of the tool. We can thus analyze these frequencies while varying the mechanical and geometrical characteristics of the cutting tool. The vibration behaviour analysis can be exploited to predict workpiece/cutting tool displacements and cutting forces of the machining system taking into account the influence of the mechanical and geometrical characteristics of the workpiece/cutting tool combination and the cutting conditions.

References

1. Mukherjee I, Ray PK (2006) A review of optimization techniques in metal cutting processes. Comput Ind Eng
2. Xiao M, Sato K (2003) The effect of tool nose radius in ultrasonic vibration cutting of hard metal. Int J Mach Tools Manuf
3. Xiao M, Karube S (2002) Analysis of chatter suppression in vibrating cutting. Int J Mach Tools Manuf
4. Thevenot V, Lionel A (2005) Influence de la position de l'outil sur le comportement dynamique en fraisage de parois minces. Mécanique & Industries
5. Bisu CF (2007) Etude des vibrations auto-entretenues en coupe Tridimensionnelle: nouvelle modélisation appliquée au tournage, Thèse doctorat
6. Bisu CF, K'Nevez JY (2008) Détermination du centre de raideur pour les machines-outils - application au tournage, Assises MUGV
7. Zghal B, Haddar M (2007) Numerical model from dynamic analysis of tool and workpiece in turning. Adv Prod Eng Manage
8. Babistsky V, Astashev V (2007) Nonlinear dynamics and control of ultrasonically assisted machining. J Vib Control
9. Ostasevicius V, Gaidys R (2010) An approach based on tool mode control for surface roughness reduction in high-frequency vibration cutting. J Sound Vib

Comparison Between Two Numerical Methods SPH/FEM and CEL by Numerical Simulation of an Impacting Water Jet

I. Ben Belgacem[1(✉)], H. Khochtali[1], L. Cheikh[1,2], E. M. Barhoumi[3], and W. Ben Salem[1,4]

[1] Laboratoire de Génie Mécanique, Ecole Nationale d'Ingénieurs de Monastir, Université de Monastir, Monastir, Tunisia
{ikrambenbelgacem, khochtalihaithem.001, wacef.bensalem}@gmail.com, lotfi.cheikh@ipeim.rnu.tn
[2] Institut Préparatoire aux Etudes d'Ingénieur de Monastir, Université de Monastir, Monastir, Tunisie
[3] Department of Electrical and Computer Engineering, College of Engineering, Dhofar University, Salalah, Oman
ebarhoumi@du.edu.om
[4] Institut Supérieur des Arts et Métiers de Mahdia, Université de Monastir, Monastir, Tunisia

Abstract. In this paper, we present a numerical simulation of a round jet impacting a polymer plate using the coupled Smoothed Particle Hydrodynamics (SPH) and Finite Element (FE) methods. Numerical results are compared with the results from other simulations carried out by other numerical method based on the Eulerian Lagrangian (CEL) formulation presented in the work of [1]. Results from the CEL method were experimentally validated in the work of [2]. A water jet with a spherical head was used at an initial speed of 570 m/s to impact on a flat plate made of Polymethyl-Methacrylate (PMMA). To simulate the entire process, the SPH method was used to model the water jet and the FE method was used for the modeling of the PMMA structure. The distribution of the pressure on the impact surface and the resulting deformation of the structure were discussed. A Numerical model was developed using ABAQUS/Explicit version 6.14. Results of the coupled SPH-FE simulation were further validated. It is demonstrated that the (CEL) method presents smoother curves compared to the SPH method. The numerical model using the SPH/FE method proves its ability to produce the entire physical processes of the impact.

Keywords: Water jet · Polymethyl-Methacrylate · Coupled Eulerian-Lagrangian · Smoothed particle hydrodynamics

1 Introduction

The impact of a high velocity jet on an obstacle is widely studied. This is due to their large applications from household appliances to space technology. They present many advantages and ease of industrialization [3]. Only a few studies have been focused

F. Chaari et al. (eds.), *Advances in Materials, Mechanics and Manufacturing*, Lecture Notes in Mechanical Engineering,
https://doi.org/10.1007/978-3-030-24247-3_7

experimentally and numerically on high-pressure water-jet taking into account the fluid/structure interaction (FSI). Simulations of IFS problems became numerous and are growing rapidly with the evolution of calculators. Some studies used the CEL approach. It has been used to simulate the FSI problem while a water jet impacting a work piece during its machining based on the orthogonal cut hypothesis [4]. The CEL method was carried out to validate experimental analyses of water jet impacting on a transparent thermoplastic polymer structure [1]; or by the use of FE analyses to simulate welding [5], tube forming [6], or else by using the LS-DYNA3D code to understand the decorating process [5]. In last decades, new studies are based on the SPH approach. It shows a good efficiency to simulate problems of fluid flow for several configurations. This method was used to treat two-dimensional interfacial flows where the influence of the density ratio variations and the role of air entrapment on loads are discussed [7]. An incompressible SPH was used to solve unsteady free-surface flows. The used method is stable and agrees with the experimental results [8]. In another study; the SPH was used to simulate the interaction between soil and water. The obtained numerical results have shown that the gross discontinuities of soil failure can be simulated easily [9]. Robustness, simplicity and relative ease of incorporating new physics are the main advantages of the SPH method [1]. In some other new studies, the coupling of the SPH method for fluid dynamics to FE method for structures was investigated. The coupled SPH- FE method was used to simulate the abrasive water jet machining. The abrasive water jet is modeled by SPH particles and the target material is modeled by FE [10]. In the medical field, a fully-coupled FSI that combines SPH and nonlinear FE method to study the coupled aortic and mitral valves structural response and the bulk intraventricular hemodynamic in a realistic left ventricle model during the entire cardiac cycle [11]. The application of the coupled SPH-FE method to fluid-structure interaction is proven by the simulation of the drop of a flexible cylinder in water [12]. In this paper, we propose a numerical study on high-speed water jet impacting a PMMA target using the SPH method for the water and the FE method for the structure. Results such as the distribution of the pressure on the impact surface and the resulting deformation of the structure will be compared to results from [1] where the CEL method was used to simulate the water jet impact. In [1] experimental results were proved by Obara et al. [2]. In this study, the numerical model is carried out under ABAQUS/Explicit code version 6.14.

2 Numerical Background

ABAQUS/explicit code version 6.14 presents the possibility to solve nonlinear physical problems using numerical methods based on Eulerian and Lagrangian approaches such CEL and SPH methods. The CEL technique combines two mesh approaches Lagrangian and Eulerian in the same analysis [1], while the SPH technique is a free mesh and based only on a Lagrangian formulation. In one hand, the CEL method is used to avoid mesh issues when running simulations contain high deformations [13]. The remaining components are meshed using the conventional Lagrangian technique. The interaction behavior between is handled by Contact definition. When performing a CEL simulation, the part dimensions of both Eulerian and Lagrangian parts should be carefully selected. The size of the Eulerian part should be selected in such a way that no material is expected to flow out of it during simulations, because as it flows out it becomes invisible to the ABAQUS solver [14]. The CEL analysis is restricted only to dynamics problems in which inertia forces are considered [14]. In ABAQUS only explicit integration of the governing equations is allowed Documentation of ABAQUS 6.14. It limits the application of the CEL analysis to short-term problems. In the dynamic explicit approach the stable time increment is usually very small, thus the long-term analysis would require millions of load increments and in consequence would take weeks or even months on powerful workstations [15]. In the other hand, the SPH presents a good alternative to resolve difficulties associated with fluid flow and structural problems presenting large deformations and free surfaces. Furthermore, for fluid dynamics problems governed by the Navier-stoke equations, the SPH method is a very powerful method. It is implemented in ABAQUS based on the one node elements formulation. ABAQUS 6.14 offers the possibility to directly define particle using the conversion to particles technique. It is an alternative based on defining conventional continuum finite element and automatically converting them to particle elements at the beginning or during the analysis Documentation of ABAQUS 6.14. The studied system is represented by a set of particles with predefined material properties and mechanical behavior low. Generated particles are interacting with each other in a domain controlled by à smoothing function. In that respect, the SPH method is quite similar to the FE method. The SPH method is based on an evolving interpolation scheme to approximate a variable field at any point in the studied domain. The value of a physical variable at a particle can be approximated by summing the contributions coming from a set of surrounding particles, for which the "kernel" function, is not zero. The SPH method has advantages relatively to other traditional numerical methods using a grid to discretize a problem. SPH can easily deal with complicated geometries and large regions of space which are, in some case, devoid of particles. The readers may refer to [16] for more details about the SPH approximation, techniques, and SPH smoothing function. In this paper, we present a comparative study of a dynamic jet water problem impacting a flat PMMA structure. A numerical model using the SPH method was developed under ABAQUS/EXPLICIT. The water is modeled by the SPH technique while the PMMA structure is modeled by the FE method. Results of this study are compared with results presented in Hsu et al. [1]. This comparison shows the effectiveness of the proposed method.

3 Problem Formulation and Numerical Study

In this study, the impact time is very short. So the heat exchange and vaporization of water are not considered [1] (Table 1).

In this study, the water is modeled using the linear Huguenot form of the Mie-Grneisen equation of state. It is a relation between the pressure and the volume of a material at a given temperature [1]. ABAQUS/Explicit provides a linear Us–Up equation of state model that can simulate incompressible viscous and inviscid laminar flow governed by the Navier-Stokes Equation of Motion [14]. Equation parameters appear in Table 2.

Table 1 Values of physical properties of water

Parameters	Symbol	Unit	Value
Density	ρ	Kg m^{-3}	0.4
Dynamic viscosity	μ	Pa s	0.45

Table 2 Parameters for the US-UP equation of water

Parameters	Symbol	Unit	Value
Reference speed of sound in water	C_0	m s^{-1}	1450
Slope of Us-Up curve	s	–	0
Grüneisen ratio of water	$\Gamma 0$	–	0

A horizontal structure of PMMA is subjected to the impact of a pure water jet which has an initial velocity of Ve = 570 m/s through a nozzle having a diameter D of 3 mm. The thickness (e) of the PMMA sheet is 5.9 mm. These geometrical parameters values are presented in Fig. 1.

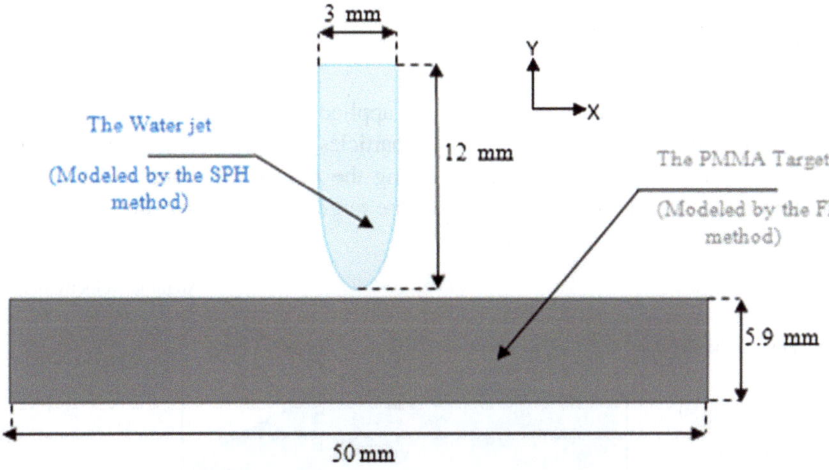

Fig. 1 problem presentation of the water jet

A coupled model has been adopted to model the target and the water using, respectively the SPH method and FEM. For the SPH method a set of points are necessary to represent the column of water. These nodes are referred to particles or pseudo particles. Starting by meshing the column of water with C3D8R (8-node linear brick, reduced integration, hourglass control) to generate "parent" elements. These "parent" elements will be after that converted to internally generated SPH particles (PC3D elements) [14]. In order to carry out a good convergence study and to highlight the behavior of the target under the first shock wave resulted from the impact of water; the refinement of the mesh has been concentrated towards the center of the target. Table 3 details the finite element model of the plate, and the water jet.

Table 3 The description of the model

Target	Water jet
Size 50 * 50 * 5.9 mm^3	**Size**: **Diameter**: 3.0 mm and Length: 12 mm
Number of nodes: 773,025	**Number of nodes**: 38,850
Number of elements: 708,396 linear hexahedral elements of type C3D8R	**Number of elements**: 36,396 linear hexahedral elements of type C3D8R converted to particles (one node elements: PC3D type)
Material property: PMMA • Density: 1170 kg/m^3 • Young modulus: 3.3 GPa • Poisson's ratio: 0.36 • Yield stress: 220 MPa	**Fluid property**: Water • Density: 1000 kg/m^3 • Dynamic viscosity: 10^{-3} Pa s • EOS type: Us–Up • Speed of sound: 1500 m/s

In SPH method, boundary conditions are applied to nodes of the parent elements and couldn't be transferred to the generated particles [14]. An initial velocity field of 570 m/s is applied to the water column during the movement along the Y-direction before impingement. The bottom surface of the target was fixed as shown in Fig. 2

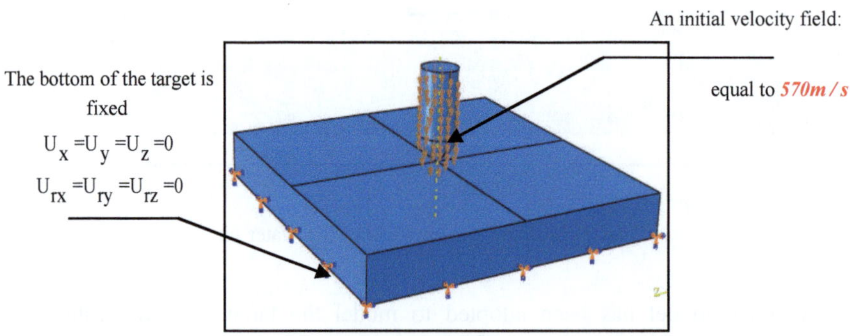

Fig. 2 Initial and boundary conditions

Moreover, general contact was used, and the type of contact domain was chosen as All*with self. Analysis time was set at 30 μs,

4 Results and Comparison

This study proposes a method using ABAQUS to simulate the dynamic process of impact by the SPH method. Simulation using the CEL method was carried out in Hsu et al. [1] where numerical results were validated by an empirical result and an experimental study presented in Mao et al. [11]. In this section, a comparison between the results coming from the two techniques CEL and SPH is presented.

4.1 Dynamical Process of the Impact

The numerical dynamical process of the impact using the SPH method under ABAQUS/EXPLICIT is presented in Fig. 3. Compared to the dynamical process of the impact presented in Hsu et al. [1], Values of the velocity are relatively close.

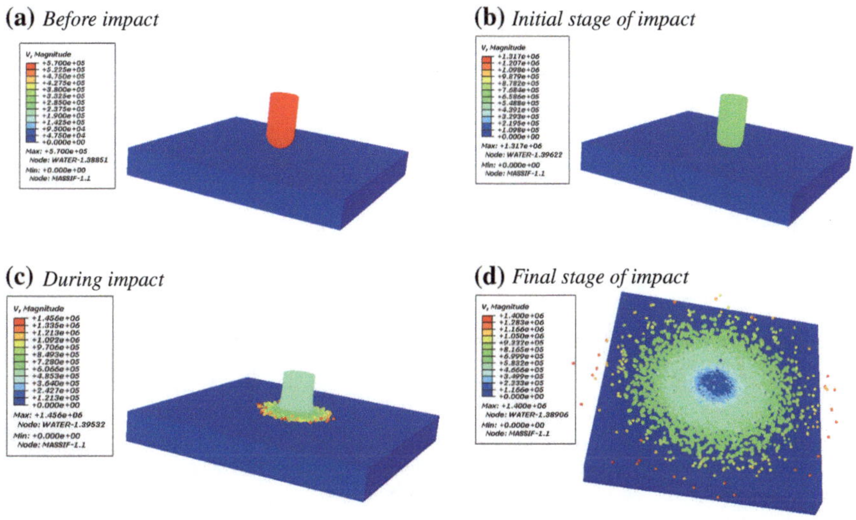

Fig. 3 The dynamical process of the SPH simulation **a** before impact, **b** initial stage of impact, **c** during the impact and **d** final stage of the of impact

At the initial stage of impact, the jet speed takes a uniform value equal to 548 m/s as shown in Fig. 2b. Using the CEL method this value is equal to 530 m/s [1]. As shown in Fig. 2c, during the impact, the velocity profile of the water jet presents a dominating value equal to 485 m/s compared to 467 m/s using the CEL method. Moreover, at the same stage, the maximal value of the velocity is equal to 934 m/s for both methods. At the final stage of the process, velocity takes 466 m/s in the neighboring areas in the center of the target as presented in Fig. 2c, while it is zero in the center. The same ascertainment presented in the CEL method, where the velocity is equal to 432.5 m/s and zero in the center [1].

4.2 The Peak Pressure and Its Duration

For a dynamic process like the impacting water jet, the initial pressure is comparable to the water hammer pressure. This pressure lasts for the time it takes the release wave, which is generated at the contact edge of the jet to reach the central point. Figure 4 presets the numerical pressure of the impact area in the center of the target from the SPH method. It shows a peak pressure on the surface just over 0.94 GPA. This value is comparable to the empirical, experimental and numerical by the CEL method results of 0.885, 0.88 and 0.9 GPa respectively presented in [1]. After peaking pressure declines and oscillates strongly between a maximal value just over 0.5 GPa and a minimal value of −0.45 GPa until the end of the impact. Results from the CEL method present a smoother oscillation before remaining stable at approximately 0.18 GPa at the end of the impact. This difference can be explained by the dispersion nature of the SPH method, contrarily to the CEL method which presents more continuity and stability. The simulation presents an agreement with CEL, experimental and empirical results from Hsu et al. [1] in the duration of the water–hammer pressure (approximately 1.5 μs).

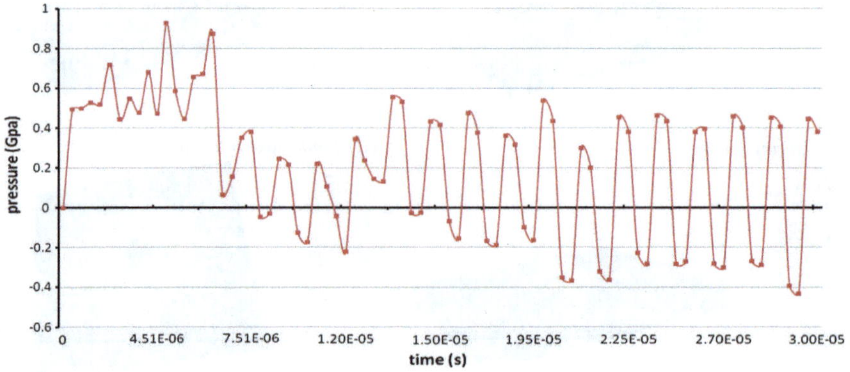

Fig. 4 The impact pressure at typical point in the plate

4.3 Von Mises Stress on the Target

Figure 5 shows the propagation of stress wave on the surface of the plate at 7.507 μs from the top view. The same evolution of the stress in both simulations: SPH and CEl. The maximum value is 2.19 and 2.2 GPa from respectively SPH and CEL simulations. The dominant value in the impact area is just over 1.28 GPa for both methods. The same coherence is observed for the minimal value of the Von Mises stress which is equal to 1.83 GPa. The Von Mises stress on the target at different intervals from a side view is presented in Fig. 6

A high pressure of 0.22 GPa occurs at the center first as shown in Fig. 6a, a dominant value of 0.128A high pressure of 0.22 GPa is observed in the impact area. Figure 6b, c show that the stress propagates on the adjacent areas to reach a dominant

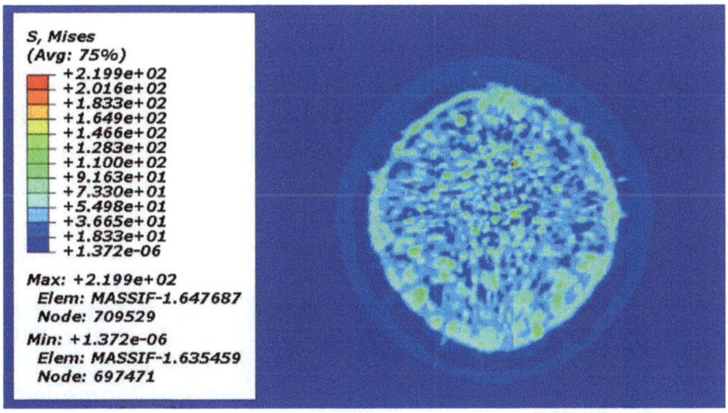

Fig. 5 The propagation of stress wave on surface of the plate at t = 7.507 μs

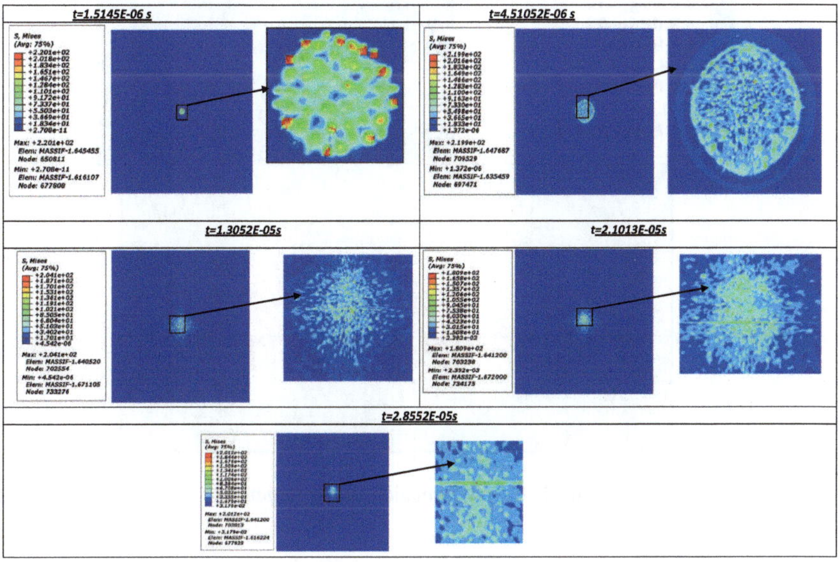

Fig. 6 Time history of Von-Misses stress in the target at different time intervals **a** t = 1.5145 μs, **b** t = 4.51052 μs, **c** t = 1.3052 μs, **d** t = 21.013 μs, and **e** t = 28.552 μs

value respectively 0.11 and 0.19 GPa. During impact, in Fig. 6d, the stress takes relatively a low value of just over 0.105 GPa. The Von Mises stress of the stagnation flow is presented in Fig. 6e. It shows a dominating value equals to 0.106 GPa. When comparing these results of the state of the Von Mises stress on the PMMA target with those from the CEL method in Hsu et al. [1], a very good coherence is observed. In the SPH simulation, the deformation of the PMMA target takes the same shape of water particles. For the CEL analyses, this shape takes a hemispherical form coming from the continuous head of the water jet.

4.4 Deformation of the Target

Figure 7 presents the deformation of the PMMA target subjected to the water jet impact. It shows a maximal displacement equal to 0.0466 mm. it has the same distribution and values presented in the CEL simulation in [1]. Figure 8 presents the profile of the depression surface for three different intervals from the impact process. It shows a good coherence with those presented in the CEL simulation.

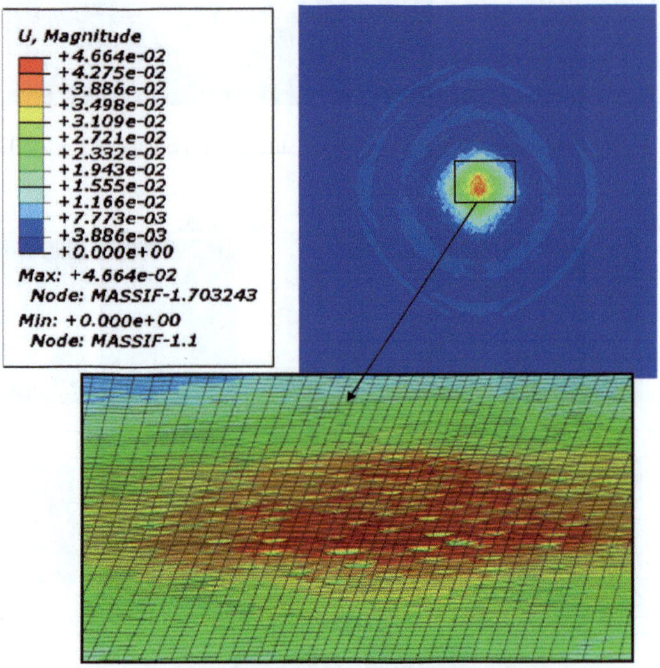

Fig. 7 The deformation of the target

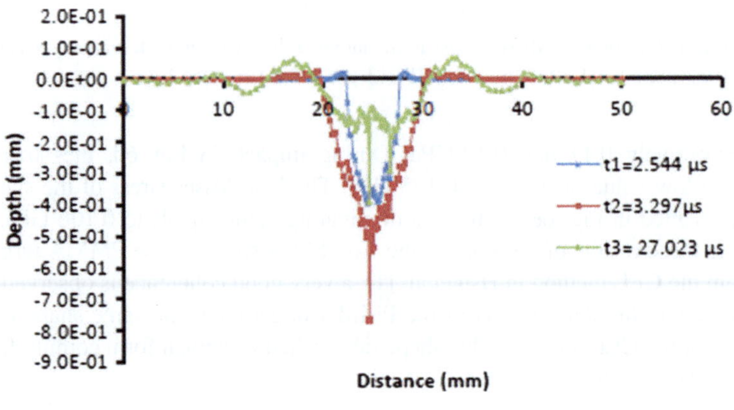

Fig. 8 Profile of the depression surface

5 Conclusion

This study presents a numerical simulation of the impact of a high-velocity water jet on a rectangular PMMA target by the SPH method for the water and an ordinary FE method for the structure. It aims to compare results with a simulation of the same problem using the CEL method presented in a previous work in [1]. Numerical models were carried out under ABAQUS; The SPH method has almost similar curves to those presented by the CEL method and experimental analyses. Moreover; the values are relatively close and have low errors. The CEL method presents smoother curves compared to the SPH method, which can be explained by the high dispersion of the water particles undergoing significant velocities.

Acknowledgements. This work is partially supported by Laboratory of Mechanical Engineering, National School of Engineers; University of Monastir. The authors also gratefully acknowledge the helpful comments and suggestions of the reviewers, which have improved the presentation.

References

1. Hsu C-Y, Liang C-C, Teng T-L, Nguyen A-T (2013) A numerical study on high-speed water jet impact. Ocean Eng 72:98–106
2. Obara T, Bourne NK, Field JE (1995) Liquid-jet impact on liquid and solid surfaces. Int J Sci Technol Friction Lubr Wear 186–187, Part 2, 388–394
3. Kaushik M, Kumar R, Humrutha G (2015) Review of computational fluid dynamics studies on jets. Am J Fluid Dyn 5(3):1–11
4. Ayed Y, Roberta C, Germaina G, Ammarab A (2016) Development of a numerical model for the understanding of the chip formation in high-pressure water-jet assisted machining. Finite Elem Anal Des 108:1–8
5. Chizari M, Al-Hassani STS, Barrett LM (2008) Experimental and numerical study of water jet spot welding. J Mater Process Technol 198(1–3):213–219
6. Chizari M, Barrett LM, Al-Hassani STS (2009) An explicit numerical modeling of the water jet tube forming. Comput Mater Sci 45(2):378–384
7. Lush PA (1991) Comparison between analytical and numerical calculations of liquid impact on elastic-plastic solid. J Mech Phys Solids 39(1):145–155
8. Manzar MT, Hosseini M (2007) An incompressible SPH method for simulation of unsteady viscoelastic free-surface flows. Int J Non-Linear Mech 42(10):1210–1223
9. Bui HH, Sako K, Fukagawa R (2007) Numerical simulation of soil–water interaction using smoothed particle hydrodynamics (SPH) method. J Terrramech 44(5):339–346
10. Jianming W, Na G, Wenjun G (2010) Abrasive waterjet machining simulation by SPH method. Int J Adv Manuf Technol 50(1–4):227–234
11. Mao W, Caballero A, McKay R, Primiano C, Sun W (2017) Fully-coupled fluid-structure interaction simulation of the aortic and mitral valves in a realistic 3D left ventricle model. PLoS One J. https://doi.org/10.1371/journal.pone.0184729
12. Groenenboom PHL, Cartwright BK (2010) Hydrodynamics and fluid-structure interaction by coupled SPH-FE method. J Hydraul Res 42:61–73
13. Documentation of ABAQUS 6.14

14. Arienti M, Hung P, Morano E, Shepherd JE (2003) A level set approach to Eulerian-Lagrangian coupling. J Comput Phys 185(1):213–251
15. Skrzat A (2012) Application of coupled Eulerian-Lagrangian approach in metal forming simulations. zeszyty naukowe politechniki rzeszowskiej, Mechanika 84(284):4–12
16. Liu MB, Liu GR (2010) Smoothed particle hydrodynamics (SPH): an overview and recent developments. Arch Comput Methods Eng 17(1):25–76

Comparison of Knife Fabricated from Tool Steel by Heat Treatment and Knife Fabricated from Structural Steel by Hard Surface Welding

Francisko Lukša(✉), Željko Domazet, Miro Bugarin, and Lovre Krstulović-Opara

Faculty of Electrical Engineering, Mechanical Engineering and Naval Architecture, University of Split, Split, Croatia
francisko.luksa@st.t-com.hr, {domazet,bugarinm, Lovre.Krstulovic-Opara}@fesb.hr

Abstract. Shears with the parallel knives are used for cutting bars to desired lengths from the mill stock. The knives are manufactured mostly from tool steel by heat treatment, while hard surface welding mainly used in repairs damaged parts. This paper presents a comparison of the ordinary knife fabrication from the tool steel X210CrW12 by heat treatment and knife fabrication from the structural steel St 52-3 by hard surface welding. Detail procedure of knife fabrication from tool steel is presented. The material X210CrW12 is received from supplier in soft annealed condition. After machining, following stress-relief annealing, preheating and austenizing, quenching, tempering and cooling on the end. The procedure of knife fabrication from the structural steel is simpler. The material St 52-3 is supplied in normalized condition. After machining and preheating, first welding is welding of ductile and high strength build-up layer, then following hard surface welding. An advantage of the fabrication by hard surface welding compared to the ordinary knives fabrication from tool steel is lower price, time of fabrication and possibility to extend total durability by several times repeating hard surface welding, but this requires skilled maintenance workers.

Keywords: Knife fabrication · Heat treatment · Hard surface welding · Cold cutting shear

1 Introduction

Shears with parallel knives are used for cutting bars to desired lengths from mill stock, Fig. 1. In very short time, three pairs of knives were broke down by spalling, Fig. 2. Length of the upper knife is 830 and lower knife 850 mm. Each knife is 50 mm thick. The height of the upper knife is 220 and lower knife 170 mm. The cross section of the lower knife is symmetrical rectangle (with sharp edges at the angle of 90°) so that all four corners can be used for the cutting. Upper knife is asymmetrical in height and can use only two sharp edges, Fig. 3.

To avoid stopping production, new knives were made from structural steel St 52-3 according to German standard DIN (Deutsches Institut für Normung) by hard surface

F. Chaari et al. (eds.), *Advances in Materials, Mechanics and Manufacturing*,
Lecture Notes in Mechanical Engineering,
https://doi.org/10.1007/978-3-030-24247-3_8

Fig. 1 Cold shear

Fig. 2 Failure of one knife

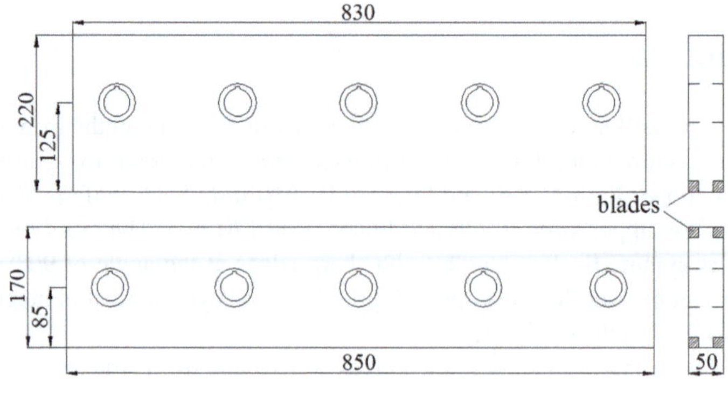

Fig. 3 Upper and lower knife

welding. This paper presents a comparison of blade fabrication from tool steel X210CrW12 according to German standard DIN by heat treatment and blade fabrication from structural steel by hard surface welding.

2 Fabrication of Knives from Steel X210CrW12

Material used for the initial knives was tool steel X210CrW12 according to German standard DIN, with hardness of the blade surface after heat treatment 59 HRc. The material brand name is OCR12SP, (SIJ Metal Ravne online catalog). X210CrW12 is ledeburitic chromium with high durability and cutting performance as well as low distortion in hardening. Chemical composition of the knife material is shown in Table 1.

Table 1 Chemical composition of steel X210CrW12

Chemical element	(%)
C	2.14
Si	0.25
Mn	0.45
Cr	12
W	0.7

2.1 Machining

Material X210CrW12 is received from supplier in soft annealed condition with maximum hardness of 25 HRC. This condition allows the steel to be easily machined and heat treated. At the beginning of the process, before milling, workpiece has dimensions 840 × 230 × 60. The first step is milling of workpiece to get dimensions 830.3 × 220.2 × 50.5, Fig. 4, which differs from final dimensions by additional 0.3, 0.2 and 0.5 mm set aside for grinding.

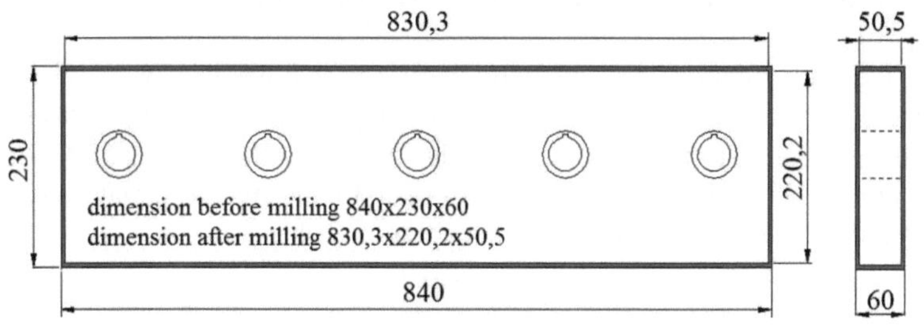

Fig. 4 Workpiece dimension before and after milling

Second step is manual grinding of corners by 0.5 mm/45°. Third step is grinding of both sides on the magnetic plate to get the thickness of 50.4 mm. Fourth step is drilling holes for the bolts and slotting. Grinding to get a final dimension follows the heat treatment.

2.2 Heat Treatment

After machining, the first step of the heat treatment is stress-relief annealing to reduce internal residual stresses in a workpiece caused by machining, without the intention to change structure and mechanical properties. This annealing is carried out by heating the workpiece on the temperature of 650 °C, holding at that temperature for one hour and slowly cooling in the furnace.

Before austenizing, due to poor conductivity, the workpiece is preheated to a temperature of 650–700 °C to equalize temperature inside and outside of the piece. Workpiece reaches the austenizing temperature much more rapidly and holding time at the austenizing temperature is reduced. The intention of preheating is to reduce change in size and risk of cracking, to avoid decarburization, grain coarsening and local overheating of the basic structure.

Preheated workpiece is transported to the furnace at the austenitizing temperature of 940–980 °C and holding time is 50–55 min.

Quenching from the austenitizing temperature can be done in oil, air or warm bath. Quenching in warm bath is used to avoid cracking of workpiece and to reduce deformation. Temperature of warm bath is 500–550 °C. Workpiece should stay in bath till equalizing temperature (cca 5 min), after which it is removed from the bath and cooled in air.

After quenching, workpiece is in a highly stressed condition. To avoid cracking, workpiece should be tempered immediately after quenching. Estimated tempering temperature is 450 °C, see Fig. 7, and duration one hour. After that, the workpiece is cooled down on room temperature.

For this material it is very important to protect the surface of tools from carburization.

Figures 5, 6 and 7 show heat treatment diagrams [1, 2].

3 Fabrication of Knives by Hard Surface Welding

Material used for fabrication blades by hard surface welding was steel St 52-3 according to DIN. Steel St 52-3 is a low carbon, high strength structural steel, suitable for welding and hard surface welding. Tensile strength of this steel is 520 N/mm^2 and the yield point is 335 N/mm^2 for thickness from 40 to 100 mm. The plate is supplied in normalized condition with chemical composition in Table 2.

3.1 Machining

At the beginning of machining, dimensions of the workpiece were 840 × 230 × 60. Workpiece was machined by milling to get final dimensions 830 × 220 × 50. Next

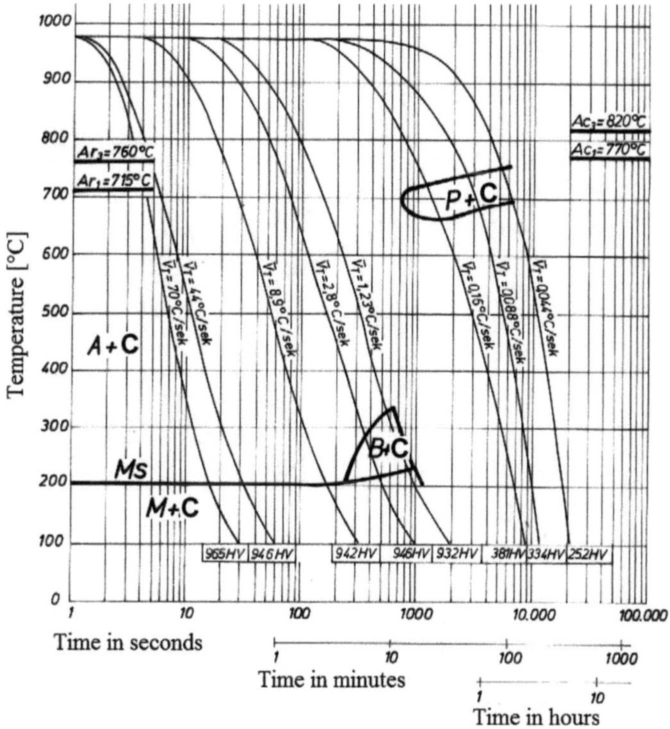

Fig. 5 Continuous cooling transformation diagram

step was drilling holes for the bolts and slotting and preparation of the grooves for welding in the corners of the workpiece, Fig. 8.

3.2 Hard Surface Welding

A sample was taken from the basic material of workpiece to confirm chemical composition by quantometer. Chemical composition determined by quantometer is shown in Table 3.

Equivalent carbon content—C_{ekv} is calculated according to (1) [3].

$$C_{ekv} = C + \frac{Mn}{6} + \frac{Ni}{15} + \frac{Mo}{4} + \frac{Cr}{5} + \frac{Cu}{13} + \frac{Si}{4} + \frac{P}{2} + \frac{V}{5} \quad (1)$$

Since $C_{ekv} = 0.43$ does not exceed value of 0.45, the workpiece was preheated to the temperature of 100 °C. Room temperature in workshop was 20 °C. Welding is done by SMAW (Shielded Metal Arc Welding) process.

First step was welding of ductile and high strength build-up layer to have sufficient strength to support surfacing layer under working conditions. This weld pass is done with electrode brand name EZ-Krom 10 R [4], classification E19 9 LR12 according to

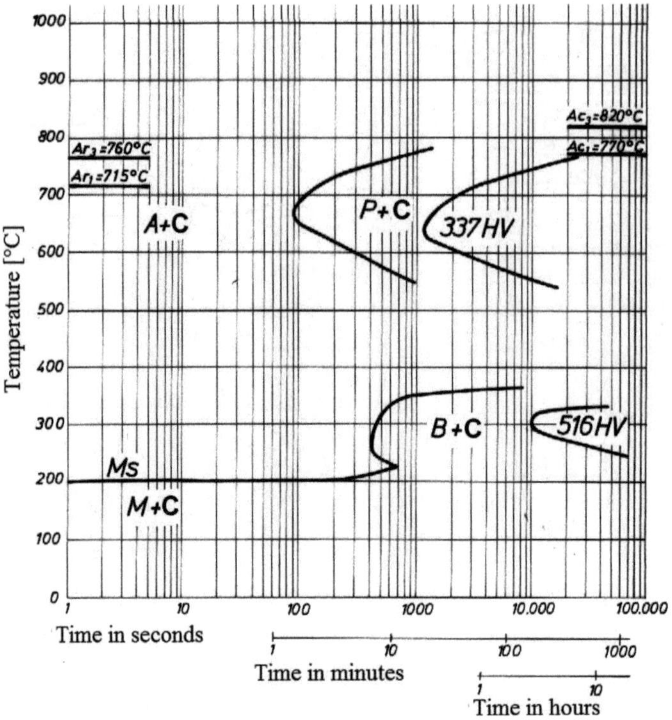

Fig. 6 Time temperature transformation diagram

European Norms EN 1600 and electrode thickness Ø3.25 mm. Approximate chemical composition and mechanical properties of the all-weld metal are shown in Table 4 [4].

Next step was welding of hard cutting edges in few weld passes with electrode brand name EZ-650 TN, [4], classification E Fe2 according to European Norms EN 14700 and electrode thickness Ø4 mm. Approximate chemical composition of the all-weld metal is show in Table 5 [4] . Estimated hardness of all-weld metal is 57-62 HRC.

Welding sequences in both cases were from the middle to the ends with back-step welding technique, Fig. 9. After welding, workpiece was covered with heat insulating material to cool down slowly. Grinding to get the final dimensions followed after hard surface welding. Hardness on the blade surface after welding was 55 HRc.

4 Comparison

Important parameters for the knives in service have been considered and compared; quality of cutting, durability, failures, possible repair and costs.

First important parameter, quality of cutting, was monitored during the production. Since the position of the cold shear is at the end of the production process of hot rolling, the appearance of the final product depends on cutting quality. Monitoring during the service showed that although hardness of blades fabricated by heat treatment

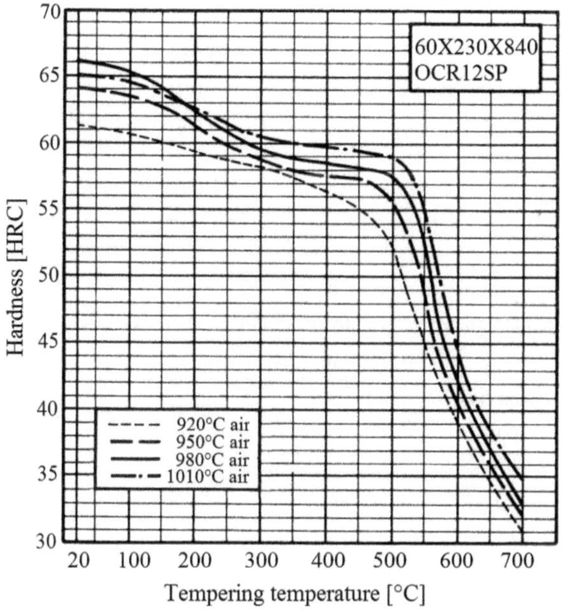

Fig. 7 Tempering temperature versus hardness of X210CrW12

Table 2 Chemical composition of steel St-52 [3]

Chemical element	(%)
C	0.2
Si	0.55
Mn	1.5
Pmax	0.035
Smax	0.035

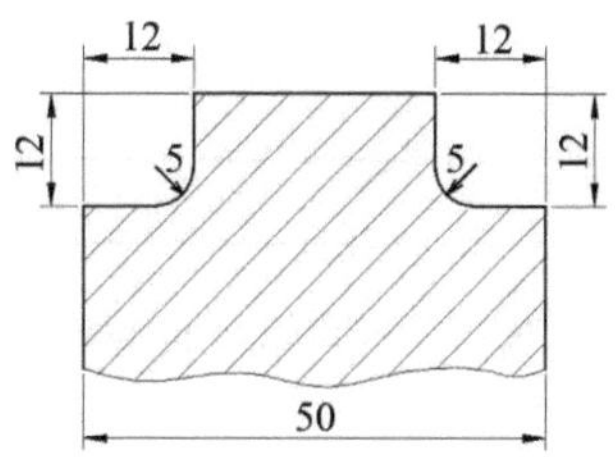

Fig. 8 Grooves for welding

was higher (59 HRc) then hardness of welded blades (55 HRc), quality of cutting was the same.

Table 3 Chemical composition of workpiece

Chemical element	(%)
C	0.19
Si	0.43
Mn	0.8
Pmax	0.01
Smax	0.01

Table 4 Approximate chemical composition and mechanical properties of all-weld metal

Chemical element	(%)	Rp0.2 (N/mm^2)	Rm (N/mm^2)	A5 (%)	KV (20 °C)
C	≤0.03	>340	540–640	>35	>55 J
Si	0.9				
Mn	0.9				
Cr	19				
Ni	10				

Table 5 Approximate chemical composition and mechanical properties of all-weld metal

Chemical element	(%)
C	0.5
Si	1.4
Mn	2
Cr	6
Ni	0.6

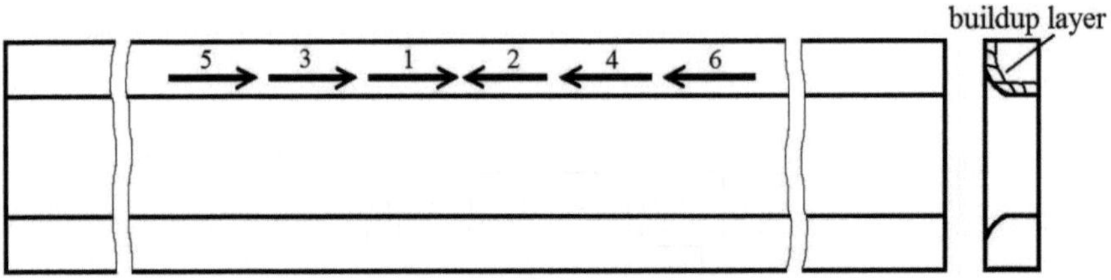

Fig. 9 Welding sequences

Second parameter, durability, was also monitored during the production. The service life of the knife made from steel X210CrW12 is around 18,000 tons of cut bars, during the period of about 3 months. The service life of the welded blades was also 18,000 tons.

Failures of blades made from steel X210CrW12 were spalling after normal service life, Fig. 2. These are large dimension blades and common failures are result of mistakes in heat treatment because of their dimensions. Inadequate heat treatment results in a lack of toughness and finally reduces service life. This kind of failure is unpredictable, which causes unpredictable stops in production and additional loss occurs due to scrap.

Welding of tool steel is complicated and requires welding skills and a lot of time. In order to reduce internal stresses and to eliminate crack formation, welding procedure of tool steel requires preheating of workpiece on high temperature, maintaining the temperature of workpiece during welding, then slow cooling and again heating after welding. Because of this, repair welding of tool steel in ordinary practice is very difficult.

Failure of knife with welded blade can be described like plastic deformation due to compressive stresses during cutting, Fig. 10. Lower knife is also subjected to wearing due to friction between rolled bars and the blade when the bars are moving in rolling direction, Fig. 1. Quality of cutting due to plastic deformation gets worse slowly and it is possible to change knives during regular production stops. After dismantling from shear, knives were welded and used in service again.

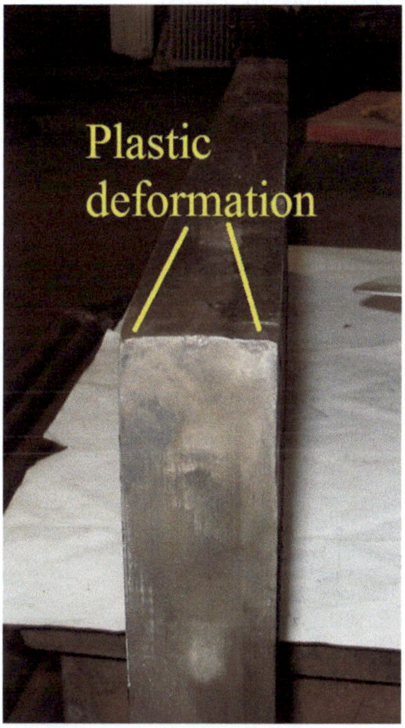

Fig. 10 Failure of blade with welded blade

Price of knives fabricated from the steel St 52-3 by hard surface welding is 70% of price of fabrication from the steel X210CrW12 by heat treatment. Costs of repair welding of knives from the steel St 52-3 are negligible.

5 Conclusion

Although the cost price of hard surface welded knife is only 30% of the cost price of knife from heat treated tool steel, possibility to repeat hard surface welding extends total durability of welded blades a few times and reduces basic costs. Time necessary to fabricate knives by hard surface welding is much shorter then fabrication from tool steel by heat treatment or buying new ones. Hard surface welding can be done in a workshop near the production line. Fabrication by hard surface welding requires skilled maintenance workers, but once they have reached requested procedure, it can be frequently repeated.

Since knives with different shapes are used on the production line in steelworks, similar welding techniques can be used to fabricate all of them.

References

1. SIJ Metal Ravne (2018) Ravne na Koroškem, Online catalog
2. Čelični materijali 1 i 2 (1990) Metalbiro. Zagreb
3. Juraga I, Živčić M, Gracin M (2001) Reparaturno zavarivanje. Publication by authors, Zagreb, 1994
4. Elektroda Zagreb (2018) Zagreb. Online catalog

The Effect of Heat Treatment on Photocatalytic Performance and Antibacterial Activity of TiO_2 Nanoparticles Prepared by Sol-Gel Method

Marwa Ben Chobba[1(✉)], Mouna Messaoud[1], Jamel Bouaziz[1], Filomena De Leo[2], and Clara Urzì[2]

[1] LCI: Laboratory of Industrial Chemistry, National School of Engineering, University of Sfax, Box 1173, 3038 Sfax, Tunisia
{marwabenchobba, Jamel.Bouaziz}@gmail.com, mouna.messaoud@enis.tn

[2] Department of Chemical, Biological, Pharmaceutical and Environmental Sciences, Viale F. Stagno d'Alcontres, 31, 98166 Messina, Italy
{fdeleo, urzicl}@unime.it

Abstract. The effect of post-heat treatment process on TiO_2 nanoparticles (NPs), prepared by hydrothermal-assisted sol-gel method, was investigated through measurement of the band gap energy, crystal size, photocatalytic activity and antibacterial efficiency of TiO_2 NPs. X-ray diffraction (XRD), Fourier-transform infrared spectroscopy (FTIR), Scanning electron microscopy (SEM) and UV-Vis spectroscopy were used to characterize treated as well as untreated TiO_2 nanoparticles. Methylene blue (MB) dye was used as a pollutant model in order to investigate the photocatalytic activity of as-prepared TiO_2 nanoparticles. The antibacterial activity of TiO_2 nanoparticles was assessed under UV-A irradiation by testing the growth inhibition of two bacterial strains: a Gram negative bacteria *Stenotrophomonas maltophilia* (*S. maltophilia*) and a Gram positive bacteria *Micrococcus luteus* (*M. luteus*). The results indicate that the crystal size increases from 13 to 20 nm for untreated (NT) and treated nanoparticles, respectively. FTIR results show that heat treatment eliminates inorganic impurity. SEM micrographs prove that annealing does not modify the morphology of nanoparticles, however, particle size increases due to calcination. In addition, post-heat treatment at 500 °C for 2 h decreases the band gap energy (from 4.35 to 3.25 eV) and consequently enhances significantly the photocatalytic activity as well as the antibacterial performance of as-synthesized TiO_2 nanoparticles.

Keywords: Heat treatment · TiO_2 · Nanoparticles · Photocatalytic activity · Antibacterial activity

F. Chaari et al. (eds.), *Advances in Materials, Mechanics and Manufacturing*, Lecture Notes in Mechanical Engineering,
https://doi.org/10.1007/978-3-030-24247-3_9

1 Introduction

Nanocrystalline titania (TiO_2) powders are of high interest because of their attractive properties such as its non-toxicity, high chemical stability, low cost and strong oxidizing power [9]. Their unique features make them a competitive candidate for many applications such as photovoltaic cells, treatment of skin tumors or skin disease, killing of bacteria, anti-fogging and self-cleaning application [1, 3, 4]. Many studies confirmed that the anatase phase of titania showed the best photocatalytic activity which makes it widely used for air purification, dangerous waste treating and water treatment. [6, 10, 13]. However, the fast recombination of photogenerated electron-hole pair is still the main weakness of anatase TiO_2 nanoparticles [7, 8]. Various studies have been developed in order to enhance the photo-response of TiO_2 nanoparticles and reduce electron-hole recombination in the photocatalytic process. It is well known that preparing methods and post-treatment conditions strongly affect the photocatalytic activity of TiO_2 because they have a critical influence on the physical and chemical properties of TiO_2 [11]. Crystal structure, particle size in addition to surface area are considered as essential factors that affect the photoactivity of TiO_2, on the other hand, heat treatment can be used to control physicochemical properties of TiO_2 [2].

In this work, the effect of heat treatment on crystalline size, particle size, structure, band gap energy, photocatalytic and antibacterial activity of sol-gel prepared TiO_2 nanoparticles have been investigated. Nanoparticles before and after heat treatment were characterized using X-ray diffraction (XRD), Fourier-transform infrared spectroscopy (FTIR), Scanning electron microscopy (SEM) and UV-Vis spectroscopy. Photo-degradation and antibacterial tests have been also carried out in order to investigate the effect of calcination on the photocatalytic performances of TiO_2 nanoparticles.

2 Materials and Methods

Sample Preparation. All the chemicals were of analytical grade and used without further purification. In order to prepare the TiO_2 nanopowder, 5 ml of ethanol, 5 ml of acetic acid and 200 μL of hydrochloric acid used as catalysts in addition to 5 ml of titanium tetra isopropoxide precursor were mixed in the indicated order. The mixture was stirred for 15 min at a constant speed of 200 rpm to obtain the TiO_2 sol. Then, the resulting sol was introduced into autoclave heated up to 243 °C and pressurized to overcome the critical point of ethanol (Tc = 243 °C, Pc = 63 bar). The sol gelation occurred after maintaining the temperature at 243 °C for 1 h. To evacuate the interstitial solvent, depressurization for 30 min down to room temperature is conducted with nitrogen gas. Finally, titanium aerogel was obtained. The obtained TiO_2 nanopowders undergo heat-treatment at 500 °C for 2 h.

Characterization of as-prepared TiO_2 nanoparticles. The structure and crystallite size of TiO_2 nanoparticles were analyzed by means of the XRD measurements which were carried out at room temperature by using (BRUKER-AXS-D8-Advance) with CuKα radiation (λ = 0.154056 nm). The average crystallite size of TiO_2 was calculated according to the Scherrer's equation using the XRD line broadening as follows:

$$D = k\lambda/\beta\cos\theta \quad (1)$$

where D is the crystallite size (Å); k = 0.89; λ is the X-ray wavelength equal to 0.154056 nm; β is the full width at half maximum intensity (FWHM) and θ—half diffraction angle.

Infrared spectra were recorded at room temperature using a Thermo Nicolet 5700 spectrophotometer. Measurements were done via transmission mode. Scanning electron microscope (SEM) was performed in order to investigate the effect of heat treatment on the morphological and particle size of nanoparticles of TiO_2. SEM images were collected by using MEB environmental QUANTA 250. Optical properties were investigated using UV-Vis spectroscopy (Optima SP-3000 plus) within the wavelength range of 200–700 nm.

Dye Photodegradation. Photocatalytic tests were carried out to evaluate the photodegradation of methylene blue (MB) dye in the presence of TiO_2 samples in aqueous solution under UV light. UV-A irradiation was performed by using three UV lamps with a maximum intensity at 365 nm. The intensity of UV radiation was found to be 0.76 mW/cm^2. The distance between the radiation source and the solution was 10 cm.

Antibacterial test. Antibacterial activity of untreated and treated TiO_2 nanoparticles suspensions (0.1 and 1% w/v) was performed under UV-A light, the test was carried out against two kinds of strains one Gram-negative bacteria (*S. maltophilia BC656*) and one Gram-positive bacteria (*Micrococcus luteus BC657*). The source of UV-irradiation was a 25 W fluorescent lamp with a maximum intensity at 365 nm. All strains were isolated and kept in the bacterial collection of the Department of Biological and Environmental Sciences (DISBA) of University of Messina, Italy. Fresh bacterial suspensions were prepared after growth in TSA medium (Tryptone Soy Agar, Oxoid) for 24–48 h and then adjusted at a final concentration of 1.5×10^7 cell/mL for Gram negative strain and 1.5×10^5 cell/mL for Gram positive strain. For the experiment, two different kinds of controls were used:

1. Bacterial suspension alone exposed under UV light and considered as a control to test the effect of UV irradiation on tested bacteria noted 'C_{uv}';
2. Another set of experiment with only bacteria not exposed to the UV was used as a control noted 'C' in order to check the viability of cells during the experimental time scale.

After 1 h of UV exposure and in order to evaluate the percentage of surviving cells, 10 μL of suspension and respective decimal dilutions were inoculated on the surface of Petri dishes containing TSA medium and then the percentage of surviving cells was estimated by counting the colonies after incubation at 28 °C for 24–48 h.

3 Results and Discussion

To determine the crystalline structure of as-prepared nanoparticles, XRD analysis was performed on both untreated and treated TiO_2 (Fig. 1). All samples were pure anatase, and no rutile phase was detected. However better crystallinity after annealing TiO_2

nanoparticles at 500 °C was observed. On the other hand, the average size of TiO_2 nanopowder increased from 13 nm to be about 20 nm for untreated and annealed sample, respectively. From these results, it could be concluded that post heat treatment enhances the crystallinity of TiO_2 nanoparticles and slightly increases their crystal size.

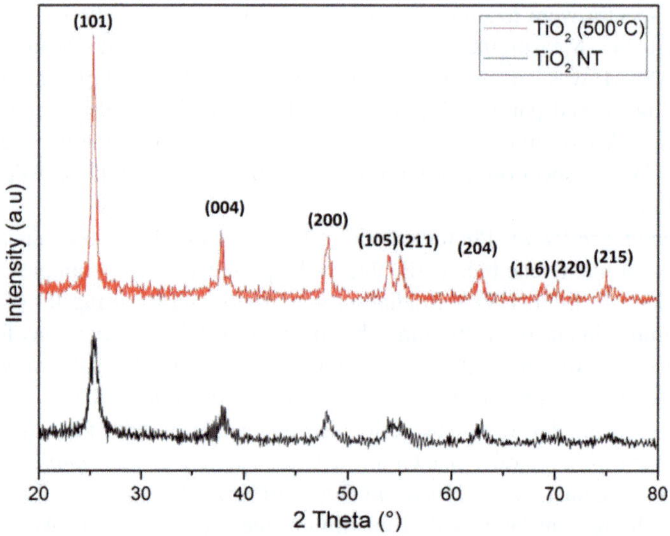

Fig. 1 XRD patterns of as-synthesized TiO_2 nanopowders

The FTIR spectrum of the as-synthesized TiO_2 nanoparticles is given in Fig. 2. The band around 437 cm^{-1} is attributed to the O-Ti-O band corresponding to the crystalline titania in the anatase form. In the case of untreated sample, picks observed at the wavenumber range between 1300 and 1800 cm^{-1}, were attributed to C = O, υ C = C and CH_3 vibration corresponding to the organic residues. A very low band, in the higher wavenumber range, between 2700 and 3800 cm^{-1} was attributed to O-H corresponding to the surface adsorbed water and C-H vibrations. However, the as mentioned peaks are not observed in the case of treated which proves that heat treatment removes organic impurities and adsorbed water which in turn could enhance the MB adsorption on the surface of TiO_2 nanoparticles and consequently ameliorate the photodegradation performances.

Scanning electron microscope (SEM) was performed in order to determine the effect of calcination on the morphology and particles size of TiO_2 nanoparticles. Figure 3 shows the SEM micrographs of TiO_2 particles and present spherical shape particles. It could be seen through the SEM analysis that TiO_2 spheres of 1 μm consist of a large amount of mono-dispersed crystallites with a size between 20 and 40 nm. Moreover, it is observed that heat treatment does not affect the morphology of nanoparticles which remain spherical after annealing. However, it is clear that particle size increased after calcination which is consistent with the XRD result.

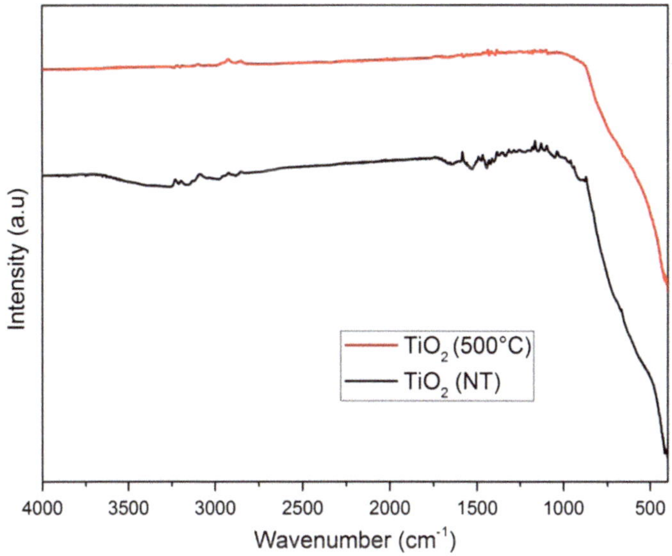

Fig. 2 FT-IR spectrum of as- prepared TiO_2 nanoparticles

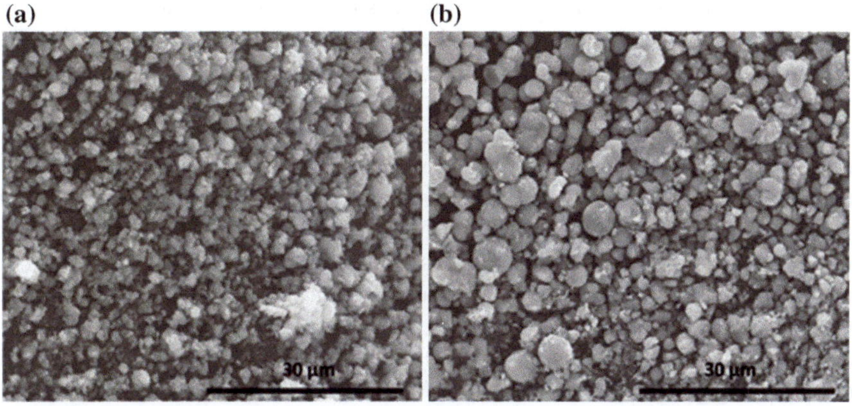

Fig. 3 SEM images of **a** not treated, **b** treated TiO_2 nanoparticles

Figure 4 shows the UV-Vis absorbance spectra of samples. The band gap energy (Eg) of TiO_2 nanoparticles before and after heat treatment process was determined by using the following equation:

$$\alpha(h\nu) = B(h\nu - Eg) \tag{2}$$

where h is Planck's constant, B is a constant dependent on the transition probability, and ν is the frequency of the radiation and α is the optical absorption coefficient.

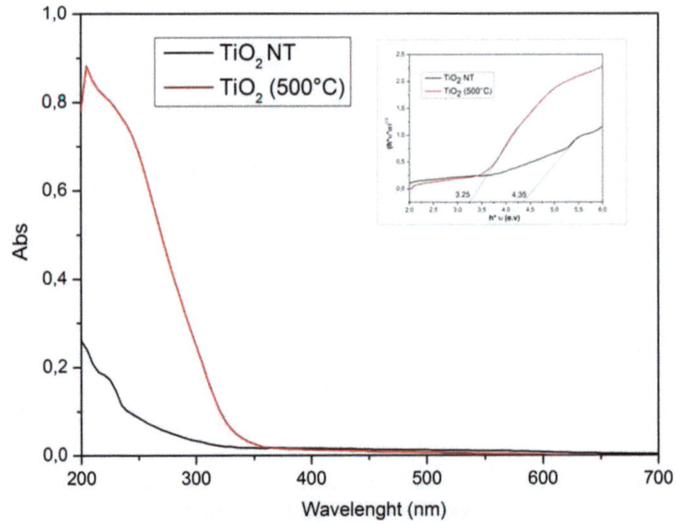

Fig. 4 Absorbance spectra untreated/treated TiO_2 nanoparticles. The optical gap of each sample has been determined from the linear fit of the Tauc's plot corresponding to each spectrum as shown as an inset

It can be clearly seen from Fig. 4 that the absorbance of UV light was significantly enhanced due to heat treatment. On the other hand, results indicate that annealing TiO_2 at 500 °C affect the optical band gap energy which decreased from 4.35 to 3.25 eV (inset of Fig. 4).

Photodegradation of MB dye was performed to investigate and compare the photocatalytic performance of treated and untreated TiO_2 nanoparticles. Figure 5 shows that not treated TiO_2 nanoparticles do not exhibit any significant photocatalytic activity by degrading only 10% of dye after 6 h. However, photocatalytic degradation rate was notably enhanced after treated nanoparticles at 500 °C. It is clear that calcination has a great effect on the photocatalytic activity of TiO_2 nanopowders. The high photocatalytic performance of annealed TiO_2 nanoparticles could be attributed to a several factors: (i) better UV-light absorption of the catalyst because of higher crystallinity and reduced band gap (ii) better adsorption of MB after the removal of organic impurities resulting from used precursors and which remain after sol-gel process (iii) the increase in the number of active sites and in the amount of photons absorbed by TiO_2.

From the results shown in Fig. 6, we can conclude that UV light has no effect on inhibiting the growth of tested bacteria. Indeed, no reduction in the percentage of survival cell was observed after 1 h of exposure under UV-A radiation. This may be due to the low intensity of the used UV lamp. Consequently, we can determine that TiO_2 nanoparticles are the only responsible for antibacterial activity and inactivation of bacterial growth for both bacteria. Calcined samples showed more efficient photocatalytic inactivation than not treated samples due to their enhanced photocatalytic activity. On the other hand, it is clear that 0.1% w/v suspension contains treated powders showed better antibacterial activity than not treated samples (1% w/v). This

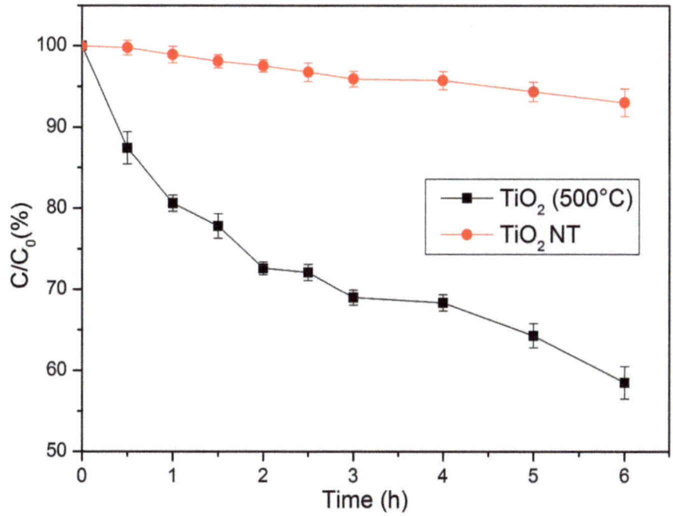

Fig. 5 Photocatalytic degradation of methylene blue dye of untreated (NT) and annealed TiO_2 nanoparticles at 500 °C for 2 h

result reveals the importance of the calcination on photocatalytic activity of TiO_2. In fact, it has been reported that the photocatalytic activity of TiO_2 is greatly affected by the calcination temperature [5, 12]. On the other hand, 1% w/v calcined suspension showed the highest antibacterial action against gram negative as well as gram positive bacteria with approximately a total inhibition after 1 hour of contact with TiO_2 nanoparticles under UV light; which exhibit their effectiveness in determining practically the complete killing of bacteria tested.

4 Conclusion

The results of this work point out that the heat treatment strongly affects the photocatalytic performance of TiO_2 nanoparticles, in particular, the photocatalytic efficiency and antibacterial activity against. Enhancing the absorbance under UV light and narrowing the band gap energy of heat-treated TiO_2 nanoparticles are the main reasons for efficient photo-response activity. The removal of organic impurities from TiO_2 NPs with heat treatment was beneficial to the studied application. The crystal size as well as particle size increased due to calcination. However, the calcination has no effect on particles morphology which remains spherical. Treated TiO_2 nanoparticles seem to be a promising candidate for many applications such as self cleaning surfaces, air purification and water treatment.

Acknowledgements. The authors gratefully acknowledge the helpful comments and suggestions of the reviewers, which have improved the presentation.

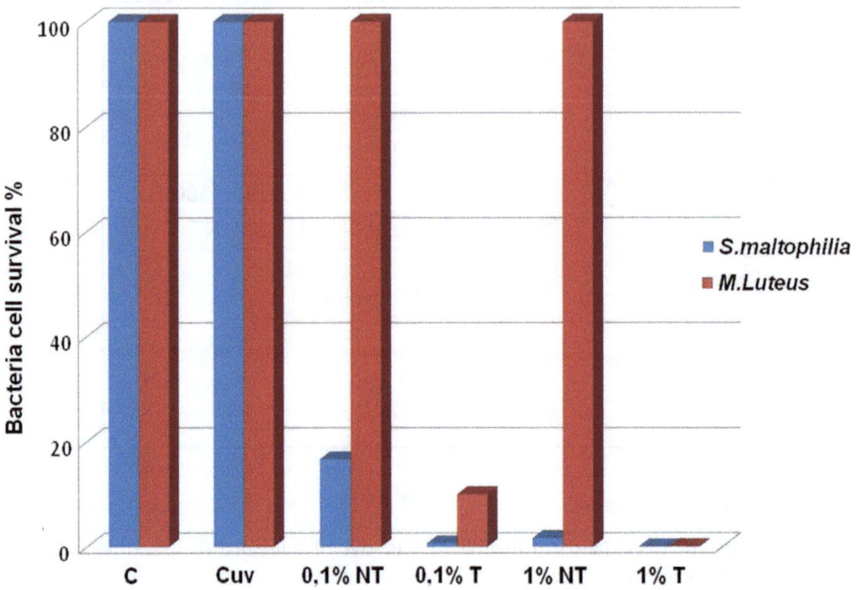

Fig. 6 Histogram showing the percentage of surviving cells of tested bacteria after contact with untreated (NT) and treated TiO_2 (T) nanoparticles suspensions, with different concentration 0.1 and 1% w/v, under UV-A light during 1 h. Controls were carried out with only bacteria

References

1. Bai Y, Mora-Seró I, De Angelis F et al (2014) Titanium dioxide nanomaterials for photovoltaic applications. Chem Rev 114:10095–10130. https://doi.org/10.1021/cr400606n
2. Behnajady MA, Alamdari ME, Modirshahla N (2013) Investigation of the effect of heat treatment process on characteristics and photocatalytic activity of TiO2-UV100 nanoparticles. Environ Prot Eng 39:33–46. https://doi.org/10.5277/EPE130103
3. Cai R, Van GM, Aw PK, Itoh K (2006) Solar-driven self-cleaning coating for a painted surface. C R Chim 9:829–835. https://doi.org/10.1016/j.crci.2005.04.007
4. Fujishima A, Rao TN, Tryk DA (2000) Titanium dioxide photocatalysis. J Photochem Photobiol C Photochem Rev 1:1–21. https://doi.org/10.1016/S1389-5567(00)00002-2
5. Guillard C, Beaugiraud B, Dutriez C et al (2002) Physicochemical properties and photocatalytic activities of TiO_2-films prepared by sol–gel methods. Appl Catal B Environ 39:331–342. https://doi.org/10.1016/S0926-3373(02)00120-0
6. Kawahara T, Ozawa T, Iwasaki M et al (2003) Photocatalytic activity of rutile-anatase coupled TiO_2 particles prepared by a dissolution-reprecipitation method. J Colloid Interface Sci 267:377–381. https://doi.org/10.1016/S0021-9797(03)00755-0
7. Mahalakshmi M, Arabindoo B, Palanichamy M, Murugesan V (2007) Preparation, Characterization, and Photocatalytic Activity of Gd^{3+} Doped TiO_2 Nanoparticles. J Nanosci Nanotechnol 7:3277–3285. https://doi.org/10.1166/jnn.2007.689
8. Rockafellow EM, Stewart LK, Jenks WS (2009) Is sulfur-doped TiO_2 an effective visible light photocatalyst for remediation? Appl Catal B Environ 91:554–562. https://doi.org/10.1016/j.apcatb.2009.06.027

9. Saif M, El-Molla SA, Aboul-Fotouh SMK et al (2014) Nanostructured $Gd3^{+}$-TiO_2 surfaces for self-cleaning application. J Mol Struct 1067:120–126. https://doi.org/10.1016/j.molstruc.2014.03.024
10. Vamathevan V, Amal R, Beydoun D et al (2002) Photocatalytic oxidation of organics in water using pure and silver-modified titanium dioxide particles. J Photochem Photobiol A Chem 148:233–245. https://doi.org/10.1016/S1010-6030(02)00049-7
11. Yu J, Yu H, Cheng B et al (2006) Enhanced photocatalytic activity of TiO_2 powder (P25) by hydrothermal treatment. J Mol Catal A Chem 253:112–118. https://doi.org/10.1016/j.molcata.2006.03.021
12. Yu J, Yu JC, Ho W, Jiang Z (2002) Effects of calcination temperature on the photocatalytic activity and photo-induced super-hydrophilicity of mesoporous TiO_2 thin films. New J Chem 26:607–613. https://doi.org/10.1039/b200964a
13. Yu JC, Yu J, Zhang L, Ho W (2002) Enhancing effects of water content and ultrasonic irradiation on the photocatalytic activity of nano-sized TiO_2 powders. J Photochem Photobiol A 148:263–271

Mechanical-Probabilistic Model of Composite Patch-Repaired Aluminum Plates Under Cyclic Loading

Houaria Errouane(✉), Khamis Hadjazi, Nadjia Deghoul, Zouaoui Sereir, and Aicha Boussoufi

Laboratory Composite Structure and Innovative Materials, Department of Marine Engineering, Faculty of Mechanical Engineering, USTO, BP 1505 El M'naouer, Oran, Algeria
herrouane@yahoo.fr

Abstract. This paper presents a probabilistic analysis approach applied to finite element analysis for modeling a cracked aluminum plate repaired with composite patches under cyclic loading. For this, it is necessary to have a mechanical model and a probabilistic model correctly representing the behavior of this type of structures. The finite element method reported in this paper to analyse the evolution of the stress intensity factor and to evaluate the effect of the composite patch on increasing the life of cracked structures. The uncertainty of the geometric characteristics and mechanical properties of the Glass/Epoxy repair patch was presented in this study. The Probabilistic method applied to finite element modeling provides another alternative medium for structural analysis of repairing aluminum plates to achieve a robust and reliable design in a more efficient manner. The Monte Carlo simulation was used in this study and the reliability in this context is defined as the probability that the stress intensity factor is less than the toughness under cyclic stress. According to this study, the most influential parameter that has a significant effect on the stress intensity factor is the thickness of the adhesive and the thickness of the patch that must be tightly controlled.

Keywords: Finite element modeling · Monte Carlo simulation · Patch repair · Probabilistic analysis · Composite material · Stress intensity factor · Fatigue · Cyclic loading

1 Introduction

The aeronautical components are subjected in service to mechanical stresses giving rise to generally multi-axial and variable amplitude stress states, which causes a phenomenon of fatigue damage. This type of structure can suffer damage or local microcracks. The damage that results is rarely singular, but it can potentially have adverse consequences on the mechanical behavior of the structure in question. An alternative to repairing these structures is to reinforce them preventively before the crack appears. Several repair techniques can be highlighted. The most successful techniques are usually the repair of damaged parts by intact plates fixed by a riveted, bolted

F. Chaari et al. (eds.), *Advances in Materials, Mechanics and Manufacturing*, Lecture Notes in Mechanical Engineering,
https://doi.org/10.1007/978-3-030-24247-3_10

(mechanical assembly), welded or glued joint. Designers of metal structures and mechanical systems are constantly confronted with fatigue problems. Cyclical solicitations can sometimes be catastrophic, even if they largely respect the levels of static sizing [1, 2]. Mechanical assembly techniques, although effective and perfectly controlled, have the disadvantage of creating areas of high concentration of stress at the connection points and expose the assemblies to high risks of degradation by corrosion. New approaches to the repair of metal structures by new materials such as composite materials, using new production techniques are currently proposed by the designers, with the primary concern is to improve the performance of aluminum structures and extend the life of the structures. Bonding offers the possibility of avoiding the different problems of mechanical assemblies, thanks to a uniform distribution of stresses on the contact surface and to the use of polymer adhesives and composite patches that are not subject to corrosion [3]. These composite patches can delay the propagation of cracks and consequently increase the life of the structures thus repaired, the asymmetry of the repaired structure, the tensioning of the component before gluing the composite patch [4, 5], imperfect gluing of the patch composite [6] or residual stresses [7], the size and properties of the material of the adhesive, the patch and the plate on the evolution of the stress intensity factor [8], the hydrothermal effect on the lifespan of aluminum plates repaired with unidirectional composite patches [9] have been examined in the recent literature. From all of the work mentioned above, it has been noted that the use of composite patches allows a significant extension of the life of the damaged structures, but in return, in this technique, the choice of shape, dimensions and the mechanical properties of the repair patch become a problem. To solve this problem, a probabilistic calculation of this technique proves essential. Consideration of uncertainty in mechanical analysis is necessary for optimal and robust design of structures. It is with this objective that probabilistic approaches for structural mechanics have been developed for several decades [10, 11]. These methods make it possible to study on the one hand the reliability of the components or systems, and on the other hand the influence of the variability of the parameters on the behavior of the component or the system. Structural reliability methods are used to assess the probability of failure of a component or mechanical system with respect to a given failure criterion [12, 13]. The database obtained by Virkler et al. [14] gave rise to different statistical analyzes for the identification of the probabilistic parameters of the propagation law used (i.e. the Paris-Erdogan law) to predict the dispersion of the propagation curves of crack obtained experimentally. Sensitivity and probabilistic analysis of the reliability of the foundation of the machine according to the variability of soil stiffness, structure and geometry and operation of the compressor was presented by Králik et al. [15]. The randomness in the loading and the environmental effects, the variability of the material and the geometric characteristics of these structures and many other "uncertainties" affecting the errors in the computational model leads to a situation in which the real behavior of the structure is different to the modeled behavior is presented by Marek P. et al. [16]. A probabilistic analysis approach applied in finite element analysis for modeling prestressed inverted T-beams with web openings structure used in building service system (mechanical, electrical, communications, and plumbing) is presented by Ming et al. [17]. Shui-Hua et al. [18] have proposed a non-intrusive stochastic finite element method for slope reliability analysis considering spatially variable shear strength parameters. They found that the non-intrusive stochastic finite element method can efficiently evaluate the slope reliability considering spatially variable shear strength parameters, which is much more efficient than the Latin hypercube sampling (LHS) method. The propagation of multi-scale uncertainty and selection of failure criterion, for the reliability analysis of composites, were investigated by Zhoua et al. [19]. They noticed that the comparison of failure criteria in reliability analysis demonstrates that the difference between the failure mode based criteria and the failure criteria without failure modes becomes more evident when the structures are in complex stress status, such as under biaxial or off-axis loadings. In general, failure theories without failure modes are preferred in reliability analysis for composites due to their efficiency. Chen et al. [20] studied, a concept of repair tolerance by defining two critical thresholds, when to repair and when to replace, to address both safety and economic issues. A probabilistic approach was applied to optimize the two thresholds in repair tolerance by assessing the probability of failure as well as minimizing the total maintenance cost. This method can be used by aircraft manufacturers and operators to support their decision-making for an optimized repair policy. A multi-scale finite element based reliability analysis is proposed for a composite structure was presented by Zhou et al. [21]. The proposed method enables both microscopic uncertainties, such as those in constituent material properties, and macroscopic uncertainties, such as ply orientation angles, to be taken into account. The proposed method will serve as a fundamental component in the development of stochastic multi-scale design method for composite structures. Probability analysis of optimal design for fatigue crack of aluminium plate repaired with a bonded composite patch is presented by Errouane et al. [22], have studied a numerical model for probabilistic analysis of optimal design of fatigue non-uniform crack growth behavior of a cracked aluminium 2024 T3 plate repaired with a bonded composite patch is investigated and they concluded that the uncertainties of the optimized patch have the largest impact on the reliability of our model. Consequently, a small variation in the manufacturing process of these properties can seriously affect the fatigue life of the fatigue life repaired aluminum 2024 T3 plate, and the four input variables: transversal modulus, thickness of adhesive, thickness and width of the patch are responsible for the largest part of fatigue life reliability. In this work, a probabilistic analysis approach applied to finite element analysis for modeling a cracked aluminum plate repaired with composite patches under cyclic loading. The probabilistic analysis approach was applied to account for the variability in dimensions and mechanical properties of the patch. Monte Carlo simulation was used in this study to analyze the effect of geometric parameter uncertainty, and the mechanical properties of the Glass/Epoxy patch.

2 Materials and Methods

A numerical model was carried out on the basis of an experimental model realized by Hosseini Toudeshky [23]. According to Fig. 1, a 2024-T3 aluminum plate containing a central crack (mode I) repaired asymmetrically. Glass/epoxy patch was modeled to estimate the evolution of crack propagation under fatigue stresses. Subsequently, an estimate of the life of this plate was validated by the experimental model. The

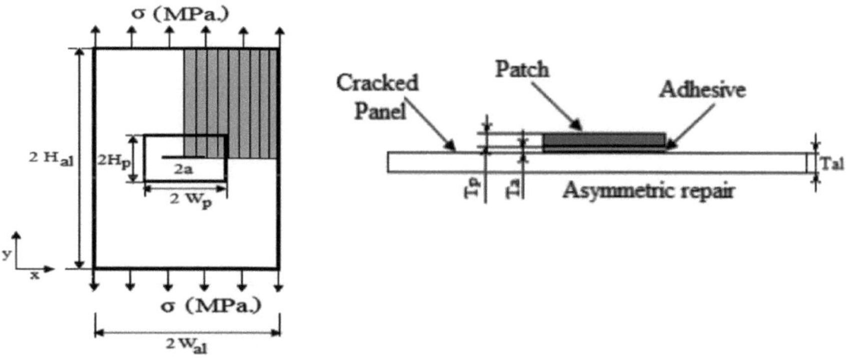

Fig. 1 Typical geometry and loading of single-side repaired panels [23]

geometric and mechanical characteristics of the aluminum plate and the patch have been grouped in Tables 1 and 2.

Table 1 Geometric characteristics of the plate, adhesive and composite patch [23]

$2W_{al}$ (mm)	$2H_{al}$ (mm)	T_{al} (mm)	$2W_P$ (mm)	$2H_P$ (mm)	T_p (mm)	Hc (mm)
50	100	2.29	35	40	0.18/plis	0.1

Table 2 Material properties of the panel, adhesive layer and patch [23]

Material	E1 (Gpa)	E2, E3 (Gpa)	ν12, ν13	ν23	G12, G13 (Gpa)	G23 (Gpa)
Aluminium 2024T3	71.3	/	0.33	/	/	/
Glass/Epoxy	50	14.5	0.33	0.33	2.56	2.24
Adhesive	1.89	/	0.33	0.33	/	/

The fatigue crack propagation analysis was carried out using the ANSYS program ware for an initial crack size a_{init} = 5 mm able to propagate to a critical final size so a_{fin} = 14 mm. The stress intensity factor is calculated at each increment of the size of this crack. According to Fig. 2, the symmetry of loading and geometry has been exploited, by the use of 1/4 of the geometry of the repaired plate. Solid 95 elements were used, for the mesh of the aluminum plate and the Adhesive film. For the composite patch, we chose Solid46 elements (Fig. 3).

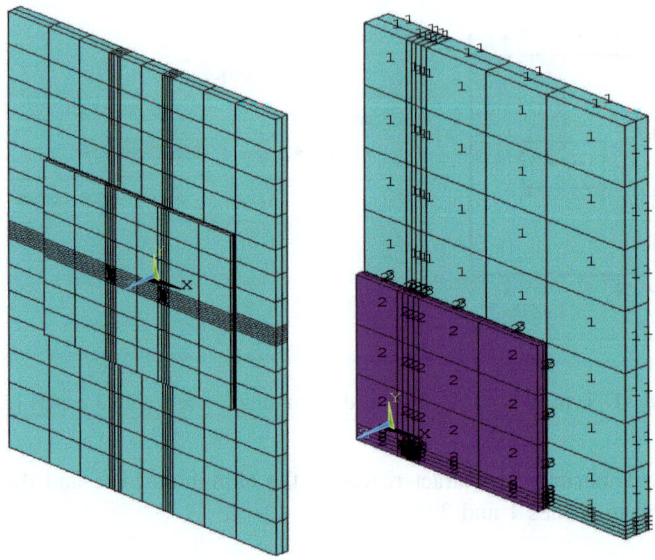

Fig. 2 Typical finite element mesh of single-side repaired panel

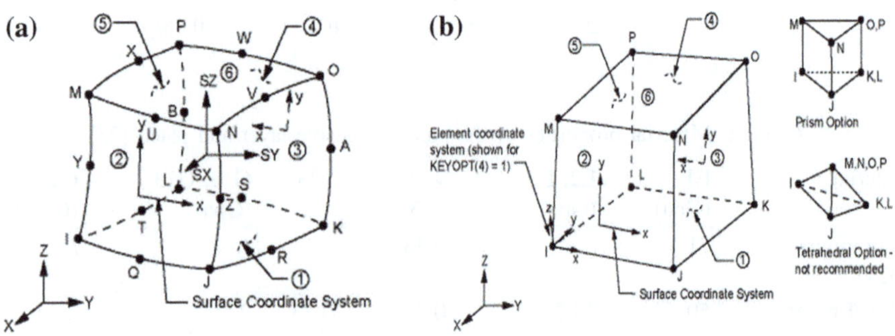

Fig. 3 Element used for meshing geometry

3 Results

3.1 Fatigue Crack Growth Life

In order to check the precision of the developed numerical model, the obtained fatigue crack growth life of the repaired thick panels using FEM is compared with experimental results performed by Hosseini and al. [23] with and without bonded patch. Figure 4 represents, the variation of crack length versus the number of load cycles for both unprepared and repaired panels with four layers of glass/epoxy patch for thin aluminum plate (Tal = 2.29 mm). The good agreement between our model and results in the literature [23] validates the developed FEM non-uniform crack growth procedure for patched and unpatched cracked aluminum plate as well as the material constants used for the Paris law.

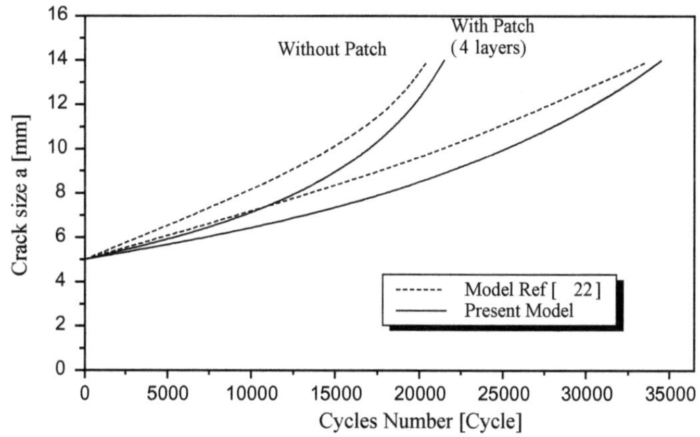

Fig. 4 Crack length versus the number of cycles for repaired and unrepaired thin aluminum plate

The Paris equation is used to relate the crack growth rate to the stress intensity factor:

$$\frac{da}{dN} = C\Delta K^m \tag{1}$$

where C and m are empirical material constants, are material constants, given by 3.2828 and 3.63×10^{-13} respectively for thin panels and 4.224 and

$$\Delta K = K_{\max} - K_{\min} \tag{2}$$

is the stress intensity factor range in fatigue loading, N is number of cycles, and da is crack extension length. The material constants in Paris equation are calculated based on the ASTM E-647 method. The required experimental crack growth data were obtained from the tests of un-repaired panels with a central crack [23].

3.2 Probability Analysis of Composite Patch

The Probabilistic Design System (PDS) analyzes a component or system involving uncertain input parameters. These parameters are defined as random input variables and are characterized by their type and the distribution of variables (mean, standard deviation). The main results of the simulation are defined as random output parameters. The mechanical parameters include both the geometry of the structure and the mechanical properties of the material. The geometric dimensions of a structure are never perfectly known. In this study, the coefficient of variation of the dimensions is fixed at 5%. The dimensions of the patch, including the width Wp, the height Hp, the thickness Tp and thickness of the film Adhesive Hc are taken as random input parameters. These quantities are therefore considered to be random and also described by a Gaussian law. In this study, the Monte Carlo MCS method is used to perform probabilistic patch analysis in the composite. The stress intensity factor KI has considered random output responses. The mechanical properties of the repair patch made of composite material, including the longitudinal modulus E11, transverse modulus E22, the poison coefficient (υ12) and the shear moduli (G12 and G23) (see Table 3).

Table 3 Random input variables of the model (a = 14 mm)

Input parameters	Means	COV % (%)	Standard deviation	Distribution
WP (mm)	17.5	5	0.875	Gaussian
HP (mm)	20	5	1	Gaussian
TP (mm)	0.72	9	0.0648	Gaussian
Hc (mm)	0.10	15	0.015	Gaussian
E11 (MPa)	50 E3	12	6000	Gaussian
E22 (MPa)	14.5 E3	12	1740	Gaussian
υ12	0.33	5	0.0165	Gaussian
G12(MPa)	2.56 E3	12	307.2	Gaussian
G23(MPa)	2.24 E3	12	268.8	Gaussian

For the probabilistic analysis, the ANSYS probabilistic design system analysis was looped through 1000 sample points considering the variations defined in the input variables, and the corresponding statistical analysis of the output parameters are given in Table 4. Figure 5 graphically represents the KI dispersion histogram for the two extreme cracks. A summary of the statistical characteristics of the output parameters is provided in Table 4, with a confidence level of 0.95. The dispersion of the input parameters causes a great variability of the output parameters.

Table 4 Probabilistic analysis results for both extreme crack sizes

	Crack a = 5 mm			Crack a = 14 mm		
	Means	SD	COV %	Means	SD	COV %
KI (MPa)	405.3	12.29	3	650.28	44.71	6.87

In this search, the value of the output parameter list in Table 4 which obtained from using the probability analysis has a different value between deterministic analyses. For example, the maximum value of KI is 405.32 $MPa.^{0.5}$ of deterministic results with probabilistic results the maximum of value of KI is 419.69 $MPa.^{0.5}$.

So from the results from Fig. 5, the approach of probabilistic analysis is relatively higher than deterministic analysis.

With respect to all geometric input parameters of 5% coefficient of variation, and for mechanical properties by 10%, the coefficient of variation of the stress intensity factor (output parameter) is too small equal to 3% for a small crack size; this coefficient was increased about 6.9% for a large crack size.

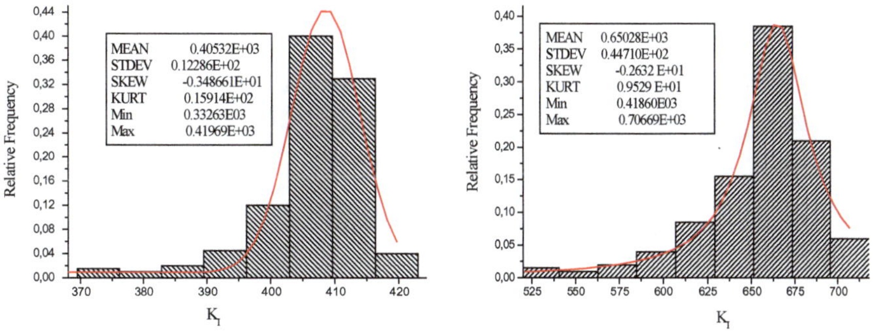

a =5mm *a* =14mm

Fig. 5 Histogram of the stress intensity factor for Initial patch

Asymmetry coefficient (skewness), from the probability distribution of a random variable of real values (Fig. 5), qualitatively shows a negative inclination for two extreme crack sizes. In the same figure, the tail on the left side of the probability density function is longer than the right side and the majority of the values are on the right side of the average. The asymmetry value for a crack of a = 5 mm and a = 14 mm is about −3.5 and −2.63 respectively. These values are important compared to the normal because it is greater than zero. Since the flattening coefficient (kurtosis), the probability distribution of a random variable with respect to the actual value for a small crack is higher (Fig. 5) represents a value of 15.91, and 9.52 for a size of the crack a = 14 mm, which means the variance represents the result of the less frequent extreme deviations, as opposed to frequent discreet size discrepancies. As shown in Fig. 5 the minimum and maximum of KI are 332.63 and 419.69 respectively, and it increases by about 26.2%.

3.3 Sensitivity Analysis of Output Variables

The evaluation of the probabilistic sensitivities is based on the correlation coefficients between all random input variables and a particular random output parameter. To plot the sensitivities of a certain random output parameter, the random input variables are separated into two groups: those that are significant (important) and those that are insignificant (not important) for the random output parameter. The sensitivity plots will only include the significant random input variables. The results of the sensitivity analysis of stress intensity factor (KI) in mode I a cracked plate of crack length a = 5 mm and a = 14 mm repaired by a composite patch are shown in Fig. 6. The sensitivities are given graphically circular (pie chart).

He can observe that 3 parameters taken into account in the simulation have a contribution to the sensitivity of KI (Fig. 6a), the thickness of the adhesive which presents 54%, the thickness of the patch 13% and the coefficient 12% Poisson's have significant effects on the performance of the composite patch repair technique. These three variables account for more than 80% of the effect on the probability of failure, compared to the other six variables that together make up the remaining parts. They are

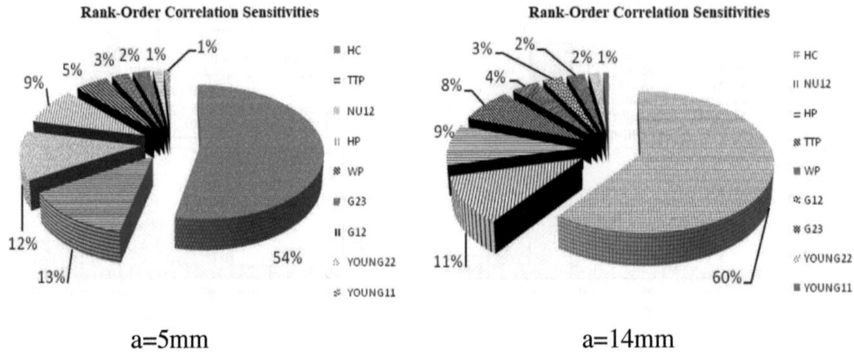

Fig. 6 Sensitivity of K_I with respect to input parameters for composite patch

followed by the height and width of the patch, the shear modulus in both directions, and Young's modulus in the longitudinal and transversal direction (Hc > Tp > υ12 > Hp > Wp > G23 > G12 > E22 > E11 …). The same observations are observed for a crack a = 14 mm, the input parameter (Hc) is the dominant input variable which represents 60% of the sensitivity of the KI, the Poisson coefficient of a percentage of 11% and the patch thickness (Tp) is only 8% when compared with a small crack. The sensitivity analysis gives us a reduction in the complexity of the problem from nine input variables to only three most dominant parameters.

3.4 Dispersion Diagrams of Output and Input Variables

The correlation coefficients enable to estimate the sensitivity of output to input variables in a qualitative manner. The problem arises when it is necessary to determine which input variable is the most significant. Such a comparison is necessary when it is desired to optimize, for example, the fatigue life of the composite material repaired structure by making changes only to the most significant input variables. A quantitative measure can be obtained by calculation of probabilistic sensitivities based on scatter plots. The scatter plots represent some interrelations between random input and output parameters as shown in Fig. 7. The correlation coefficient chosen in this work is named the Spearman rank order correlation coefficient. In particular, a qualitative sensitivity measure of output parameter Siwith respect to input parameter *Ri* through the Spearman coefficient can be calculated

$$r_s = \frac{\sum_i^n (R_i - \overline{R})(S_i - \overline{S})}{\sqrt{\sum_i^n (R_i - \overline{R})^2}\sqrt{\sum_i^n (S_i - \overline{S})^2}} \tag{3}$$

where Ri, is rank of input parameters within the set of observations $[x1, x2, x3, \ldots, xn]^T$, Si is rank of output parameters within the set of observations $[y1, y2, y3, \ldots, xn]^T$, are average ranks of the parameters $\overline{R}$ and $\overline{S}$ respectively R, S [21].

It is observed that there is a non-linear relationship between random input Hc and output KI parameters (Fig. 7). The relevant results are collected and uncorrelated case

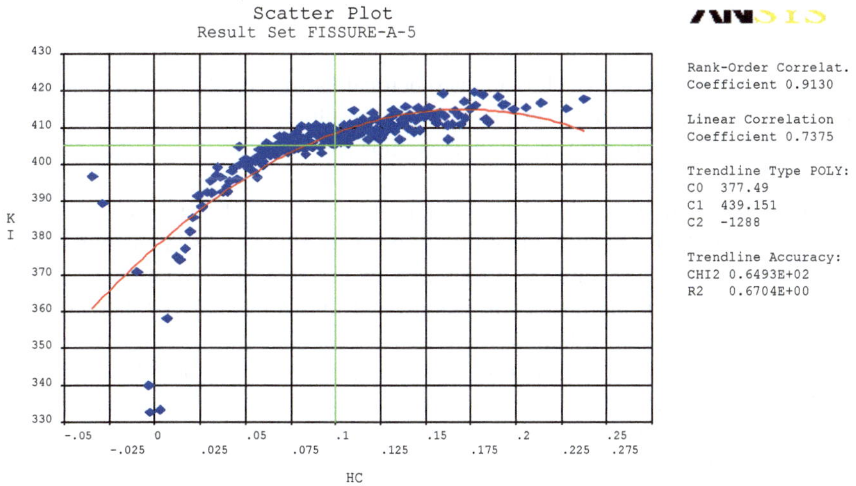

Fig. 7 Spearman Correlation Coefficient Between thickness of adhesive Hc and stress intensity factor KI for a = 5 mm

in Table 5 being equal to 1% of the corresponding mean value. The outcome of computations shows that one model parameter (Hc) is strongly correlated (nearly equal ±1) with the output parameter (KI), thus they slightly deviate from trend lines as shown, e.g., in Table 5.

Table 5 Spearman correlation coefficient Rank order between input variables and output parameters

Variables	Coefficient Spearman ordre de rang KI	
	r	$r^2 * 100$
HP	0.153	2.3
WP	−0.085	0.7
TP	−0.225	5.0
Hc	**0.913**	**83.3**
E11	0.017	0.02
E22	−0.023	0.05
G12	−0.042	0.17
G23	0.044	0.19
υ12	−0.203	4.1

3.5 The Sensitivity the Geometric Ratio α (Wp/Hp) on the KI

Consideration of uncertainty in a mechanical analysis is necessary for the optimal and robust design of structures. It is with this objective that the shape of the repair patch was tested as a function of the geometrical ratio α in order to study the influence of the variability of the shape of the patch on the mechanical behavior of the repaired structure. In this part, we have chosen 3 geometrical ratios α [0.65, 1 and 1.5] as shown in Fig. 7. Table 4 groups together the statistical results of the output parameters as a function of geometrical ratio α, with a confidence level of 0, 95 a crack length a = 14 mm.

Figures 9 and 10 show the relative density function and the curves of the cumulative function of the probability of failure (dispersion) of the stress intensity factor as a function of the geometric ratio α, a variation in the standard deviation leads to an increase or decrease in the dispersion margin and thus a decrease or increase in the slope of the probability distribution at a constant position.

In Figs. 8 and 9, according to Table 6, the average value of the KI density varies from 1.78, 1 and 3.21% of a reference value (Table 3) for α equal to 0.65, 1 and, 1.5 respectively. A reduction of 3.21% of the average of KI is obtained for α = 1.5. A variation of about 0.24, 0.15 and 2.12% is obtained in the standard deviation of a reference value for α equal to 0.65, 1 and 1.5 respectively.

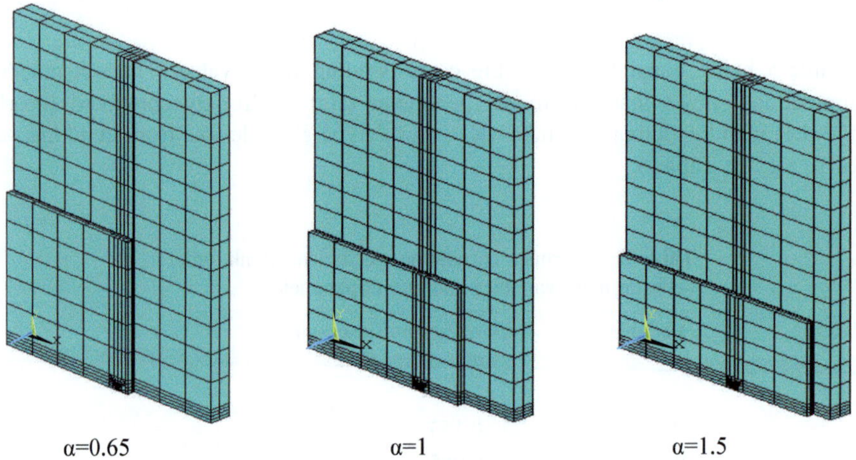

Fig. 8 Numerical model according to the geometric ratio α

A variation in the average KI value of the relative density causes a variation in the cumulative probability of failure distributions (Fig. 10) especially if the average relative density is shifted to low values. In this case, an increase in the probability of failure (52%) was observed for the geometric ratio α = 0.65, for a geometric ratio α = 1, a probability of failure has been increased to 82%.

From this study, it can be distinguished that the geometric ratio α = 1, is the suitable ratio which gives an optimal shape of the repair patch as the coefficient of variation of the mean and the standard deviation from the reference value is very small compared to other reports.

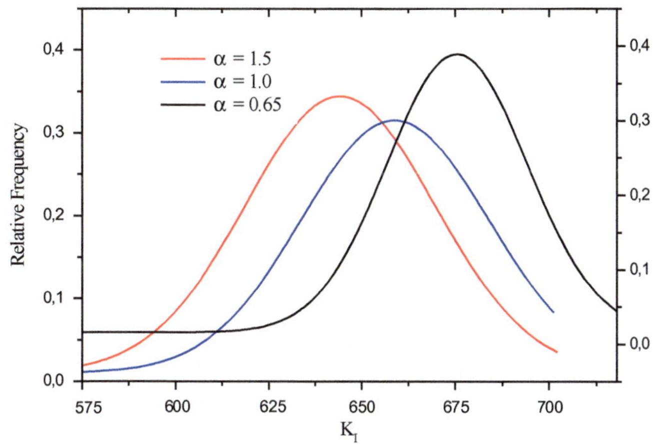

Fig. 9 Relative density of the stress intensity factor as a function of the geometric ratio α

Table 6 Probabilistic analysis results of the geometrical ratio α for a = 14 mm

	α = 0.65			α = 1			α = 1.5		
	Means	SD	COV %	Means	SD	COV %	Means	SD	COV %
KI (MPa)	661.9	44.8	6.77	643.4	44.78	6.96	650.28	44.71	6.87

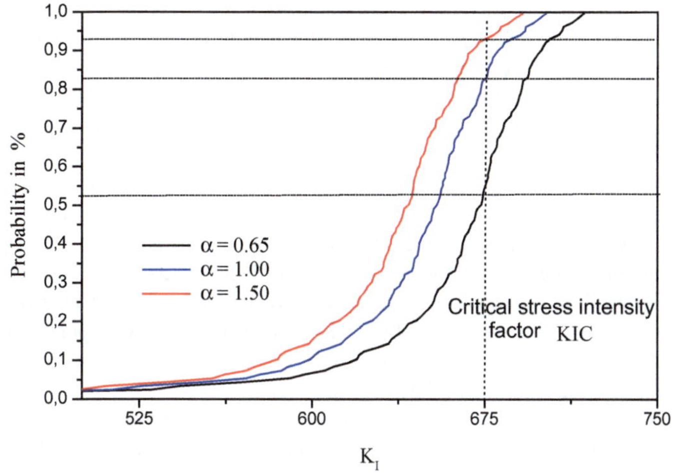

Fig. 10 Curve of the cumulative function of the stress intensity factor for different geometrical ratio α

4 Discussion and Conclusion

This work proposes a mechano-probabilistic study based on the Monte Carlo method for solving the crack propagation problems of aluminum plate repaired with composite patches under a cyclic loading modeled by the finite element method. In account uncertainties related to the geometrical parameters and the mechanical characteristics of the patch of repair. Reliability in this context is defined as the probability that the stress intensity factor is less than the toughness (KIC) under cyclic stress. According to the analysis results, three parameters taken into account in the simulation have a contribution to KI sensitivity, adhesive thickness, patch thickness, and Poisson's ratio. The variation of the shape of the patch has a very significant effect on the sensitivity of the KI. The square shape of the repair patch (geometric ratio $\alpha = 1.0$) is the optimal form in the composite material repair concept. The knowledge of the influence of each parameter of the repair patch which directly influences the propagation of the crack and above all the aeronautical structures allow us to use the best approximations which accelerate the computation time while keeping an acceptable precision, and we quantify the error made during the simplifications chosen.

References

1. Schütz W (1996) A history of fatigue. Eng Fract Mech 54(2):263–300
2. Schijve J (2003) Fatigue of structures and materials in the 20th century and the state of the art. Int J Fatigue 25(8):679–702
3. Bruneaux MA (2004) Durabilité des assemblages colles: modélisation mécanique et physico-chimique. Thèse doctorat. Ecole nationale des ponts et Chaussées. Spécialité: Structures et Matériaux. University Degli Studi di
4. Chung KH, Yang WH (2003) A study of the fatigue crack growth behaviour of thick aluminium panels repaired with a composite patch. Compos Struct 60:1–7
5. Hosseini Toudeshky H, Sadeghi G, Daghyani HR (2005) Experimental fatigue crack growth and crack-front shape analysis of asymmetric repaired aluminium panels with glass-epoxy composite patches. Compos Struct 71:401–406
6. Bassetti A (2001) Lamelles précontraintes en fibres de carbone pour le renforcement de ponts rivetés endommagés par fatigue. Thèse, EPFL, Lausanne, Switzerland
7. Naboulsi S, Mall S (1997) Fatigue growth of adhesively repaired panel using perfectly and imperfectly composite patchs. Theoret Appl Fract Mech 28:13–28
8. Sabelkin V, Mall S, Hansen MA, Vanderwaker RM, Derriso M (2007) Investigation into cracked aluminium plate repaired with bonded composite patch. Compos Struct 79:55–66
9. Ait Yala A, Megueni A (2009) Optimization of composite patches repairs with the design of experiments method. Mater Des 30(2009):200–205
10. Ergun E, Getiren TS, Muzaffer T (2010) Fatigue and fracture analysis of aluminum plate with composite patches under the hygrothermal effect. Compos Struct 92:2622–2631
11. Andrieu-Renaud C, Sudret S, Lemaire M (2004) The PHI2 method: a way to compute time-variant reliability. Reliab Eng Syst Saf 84(1):75–86
12. Muscolino G, Ricciardi G, Impollonia N (1999) Improved dynamic analysis of structures with mechanical uncertainties under deterministic input. Probab Eng Mech 15:199–212
13. Ditlevsen O, Madsen HO. Structural reliability methods. Wiley, Chichester

14. Virkler DA, Hillberry BM, Goel PK (1980) The statistical nature of fatigue crack propagation. Trans ASME J Eng Mater Technol 101:148–153
15. Králik J, Králik J Jr (2009) Probability and Sensitivity Analysis of Machine Foundation and Soil Interaction. Appl Comput Mech 3:87–100
16. Marek P, Brozzetti J, Gustar M. 2001 Probabilistic assessment of structures using Monte Carlo simulation, background, exercises and software. ITAM Academy of Sciences of the Czech Republic, GLOS s.r.o., Semily Czech Republic, Praha
17. Ming T, Cheng Hock T, Zahrizan bin Z, Ideris bin Z (2012) Application of probabilistic analysis in finite element modeling of prestressed inverted T-beam with web openings. Res J Appl Sci Eng Technol 4(4):350–366
18. Shui-Hua J, Dian-Qing L, Zhou CB (2014) Slope reliability analysis considering spatially variable shear strength parameters using a non-intrusive stochastic finite element method. Eng Geol 168:120–128. https://doi.org/10.1016/j.enggeo.2013.11.006
19. Zhoua XY, Goslinga PD, Kaczmarczyk L, Pearce CJ (2016) Exploiting the benefits of multi-scale analysis in reliability analysis for composite structures. Compos Struct 155:197–212. https://doi.org/10.1016/j.compstruct.2016.08.015
20. Chen X, Ren H, Bil C (2014) Repair tolerance analysis for composite structures using probabilistic methodologies. J Air Craft 51(6):1997–2004. https://doi.org/10.2514/1.c032635
21. Zhou X-Y, Gosling PD, Ullah Z, Kaczmarczyk L, Pearce CJ (2017) Stochastic multi-scale finite element based reliability analysis for laminated composite structures. Appl Math Model 45:457–473
22. Errouane H, Deghoul N, Sereir Z, Chateauneuf A (2017) Probability analysis of optimal design for fatigue crack of aluminium plate repaired with bonded composite patch. Struct Eng Mech 61(3):325–334
23. Hosseini-Toudeshky H, Mohammadi B, Sadeghi G, Daghyani HR (2007) Numerical and experimental fatigue crack growth analysis in mode-I for repaired aluminum panels using composite material. Compos Part A 38:1141–1148

Design and Simulation of a Low Cost Mini Solar Concentrator

Ahmed Ridha El Ouederni[1,2(✉)], Achref Wahabi[1], and Hacen Dhahri[1]

[1] Laboratoire d'Etudes des Systèmes Thermiques et Energétiques, Ecole Nationale d'ingénieurs de Monastir, Rue Ibn El Jazzar, 5019 Monastir, Tunisia
ridha_ouederni@yahoo.fr

[2] Laboratoire de recherche: "Caractérisations, Applications et Modélisation de Matériaux" LR18ES08, Faculté des Sciences de Tunis, Université de Manar, Campus Universitaire Le Belvédère, 1060 Tunis, Tunisie

Abstract. In this paper, a design of a mini parabolic concentrator is studied. The interior surface is covered with a reflective layer and equipped with a cylindrical receiver in its focal position. To validate the sizing of the main parts of our system and to make our mechanism resistant against external forces, we used the numerical simulation carried out on. Currently, solar concentration technologies present opportunities for commercial exploitation. These technologies are based on collectors that concentrate solar radiation and heat a heat transfer fluid at high temperature. Hence the idea of designing and building a parabolic solar concentrator which is composed of a reflective surface that concentrates the incident radiation to its focus. Its size obviously depends on the solar power required. This type of concentrator has the highest thermal efficiency which is around 75%. Our system consists mainly of a dish, a cylindrical receiver placed in the focal position, a recessed reflector on a nacelle, a counterweight, a support supported by a pylon and a frame fixed to the ground. The system is composed of 2 main classes. It has two degrees of freedom of rotation: class 1 to adjust the azimuth angle and class 2 for the angle of elevation.

Keywords: Parabolic · Concentrator · Solar · Energy · Concentrator · Design · Simulation

1 Introduction

According to studies by some research institutes, the energy consumption of developing countries will increase very significantly, especially after the globalization of trade and exchanges between states. The latest oil crisis and the disaster at the Chernobyl nuclear power plant have stimulated much research and development of new sources of renewable and particularly solar energy. Currently solar concentrating technologies present opportunities for commercial exploitation. These technologies are based on collectors that concentrate solar radiation and heat a heat transfer fluid at high temperature [1, 2]. Hence the idea of making and realizing a parabolic solar concentrator.

F. Chaari et al. (eds.), *Advances in Materials, Mechanics and Manufacturing*,
Lecture Notes in Mechanical Engineering,
https://doi.org/10.1007/978-3-030-24247-3_11

The parabolic concentrator is composed of a reflective surface that concentrates the incident radiation to its focus. Its size obviously depends on the solar power required. This type of concentrator has the highest thermal efficiency which is of the order of 75% [3].

As an order of magnitude, it can be considered that a concentrator with a diameter of 8.5 m can provide up to 10 KW of electricity under an insolation of 800 W/m^2. The concentrator has two degrees of freedom to effectively track the sun. This tracking is done quite autonomously, unlike parabolic trough sensors [4].

2 Model and Simulation Method

Physical model

A parabolic sensor is a concentration sensor that uses a dish-shaped reflector and focuses the sun's rays into a focal point (Fig. 2) [5]. Our system consists mainly of a dish, a cylindrical receiver placed in the focal position, a recessed reflector on a nacelle, a counterweight, a support supported by a pylon and a frame fixed to the ground. The system is composed of 3 equivalence classes [Table 1]. In addition, it has two degrees of freedom of rotation: a horizontal between class 0 and class 1 to adjust the azimuth angle and another vertical between class 1 and class 2 for the angle of elevation (Fig. 1).

Table 1 Equivalence class in the system

Class number	Components
0	Wheels; built; pylon; guide; rolling bearing; screw; nuts
1	Support; 2 rolling bearings; screw; nuts
2	Nacelle; parabola; lid; absorber (receiver); counter weight; stem

Rotation guidance in azimuth: The solution adopted is the interposition of rolling elements, and more specifically the mounting of a support, a guide and a rolling bearing bearings and which suppress the degrees of freedom in translation.

a. Rotation guidance in azimuth: It is performed by the implantation of two rolling bearings.
b. Rotation guidance in elevation: It is performed by the implantation of two rolling bearings. It is the same solution existed to guide the mechanism in rotation in azimuth.

The physical model used in our case is the simulation of masterpieces on (SOLIDWORKS Simulation). The geometric parameters of the parabolic concentrator are illustrated in Tables 2 and 3.

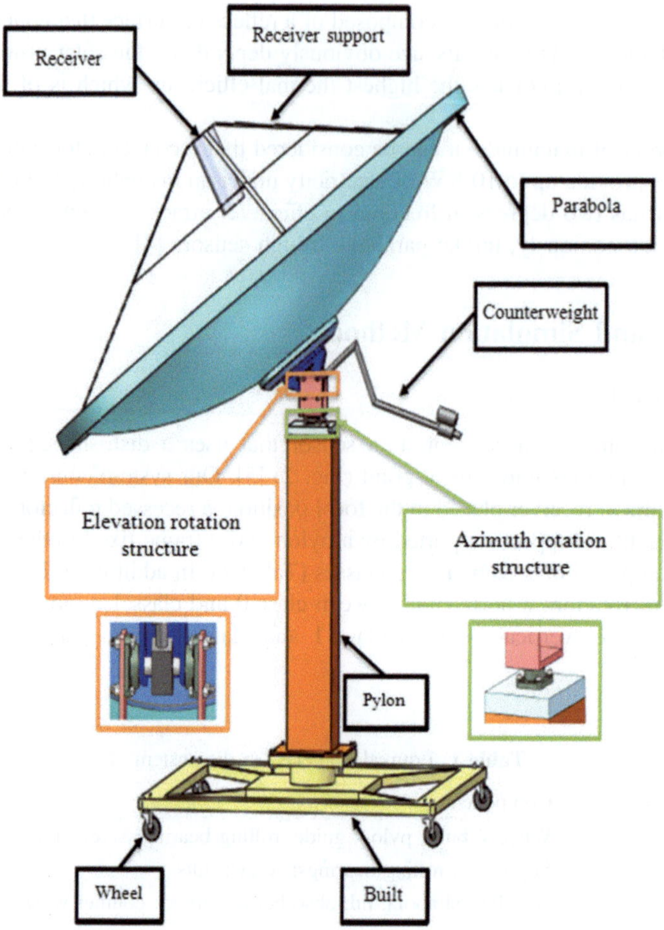

Fig. 1 Perspective of the solar parabolic concentrator

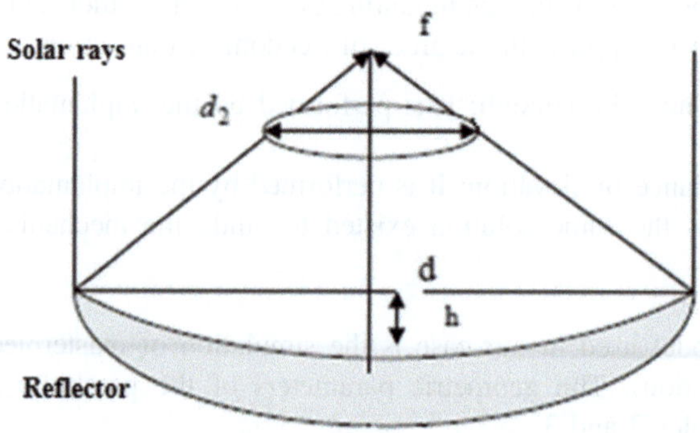

Fig. 2 Schematic of the parabolic concentrator in the direction of the circle

Table 2 Reflector parameters

Geometric parameters	Value	Unit
Diameter of the opening "d"	3	m
Depth "h"	0.5	m
Focal distance "f"	1.125	m
Opening area "A_a"	7.07	m^2

Table 3 Receiver parameters

Geometric parameters	Value	Unit
Diameter of the opening "d_2"	0.16	m
Capturing surface "A_r"	0.02	m

3 Results of Numerical Simulation

3.1 Results

Before performing the simulation of our mechanism, it is necessary to calculate the forces applied on the concentrator.

The concentrator is subject to two forces:

- The weight P of the set "2" exerted on the other structures.
- The force applied by the wind F exerted on the assembly "2".

Set "2" is essentially composed of a parabola; a nacelle; a counter weight; a rod.

$$P = M(\text{set}\,2)\,.\,G$$

The set "2" is composed of four parts that will exert the effort P on the other mechanism structures whose masses are as follows:

Mass of the dish: M1 = 125 kg; Mass of the nacelle: M2 = 7.5 kg; Weight of counterweight: M3 = 13 kg.
Mass of the empty absorber and its mounting bracket: M4 = 6.5 kg so the weight of the set "2" is:

P = (125 + 7.5 + 13 + 6.5) g

$$\rightarrow \mathbf{P = 1491.12\,N}$$

We also have:

$$F = 1/2.C_x\,.\,S\,.\,V^2\,.\,\rho$$

The force applied by the wind F exerted on the assembly “2”.

$$\mathbf{F} = 1/2\mathbf{C_x}\,\mathbf{S}\mathbf{V}^2\,\boldsymbol{\rho}$$

Cx: constant = 1.4 for a parabolic surface [2].
S: the surface, or S = $\prod R^2 = 7\ m^2$
V: wind speed = 120 km/h = 33.33 m/s;
ρ: density of the air = 1.22 $kg/m^2 \rightarrow \mathbf{F} = 6641$ N.

3.2 Numerical Simulation

Numerical simulation is performed on (SOLIDWORKS Simulation). This simulation was used to validate the sizing of the master parts of the technical system.

The resistance criterion adopted in this study is defined by a comparison between the Von Mises stress and the elastic limit of the material affected by a safety factor. At the same time, the maximum elastic deformation must be negligible [6].

$$\sigma VM \leq \sigma ad$$

With:

$$\sigma ad = Re/s$$

Re: the elastic limit of the material
s: the safety factor = 1.5 [5].

The material used for the two masterpieces (guide and pylon) is a structural steel that has the following mechanical properties (Table 4).

Table 4 The mechanical characteristics of the material “1.0035 (S 185)”

1.0035 (S 185)	
Elasticity limit	1.75×10^8 Pa
Traction limit	2.9×10^8 Pa
Young’s module	2.1×10^{11} Pa
Hear modulus	7.9×10^{10} Pa
Density	7800 kg/m^3
Poisson coefficient	0.28

3.3 Validation of the Guide

3.3.1 Model

See Fig. 3.

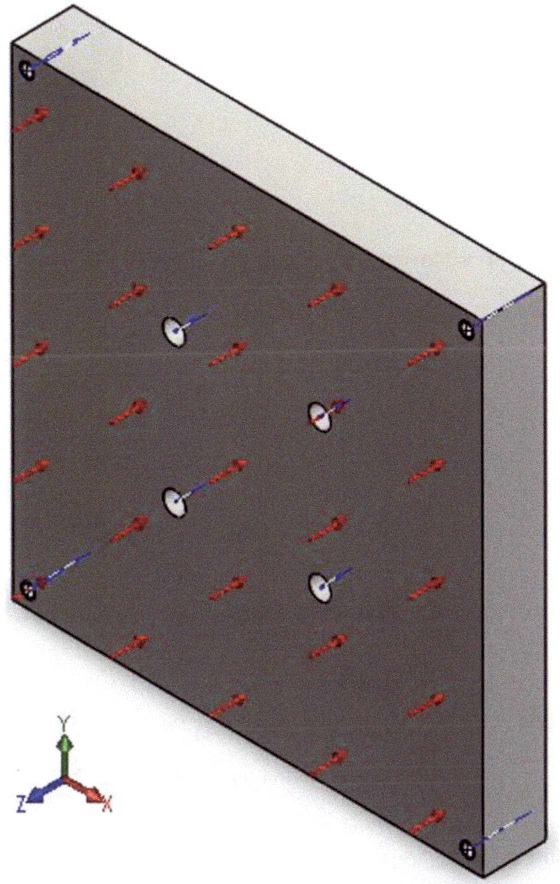

Fig. 3 Guide template

3.3.2 Simulation

See Fig. 4.

3.4 Validation of the Pylon

3.4.1 Model

See Fig. 5.

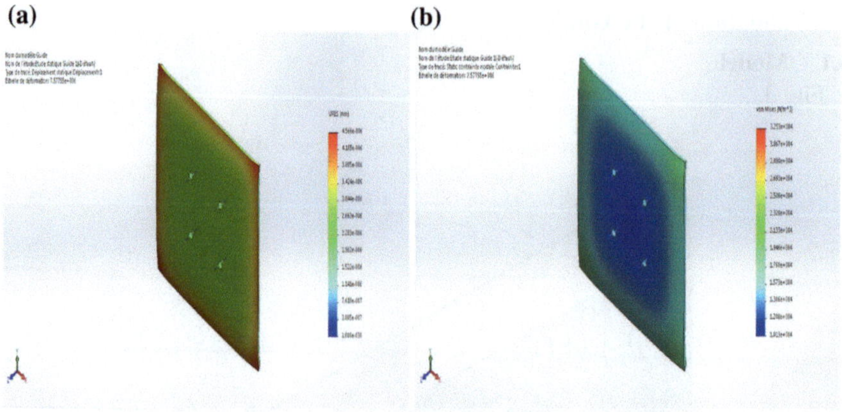

Fig. 4 **a** Moving the guide, **b** Constraint of Von Mises

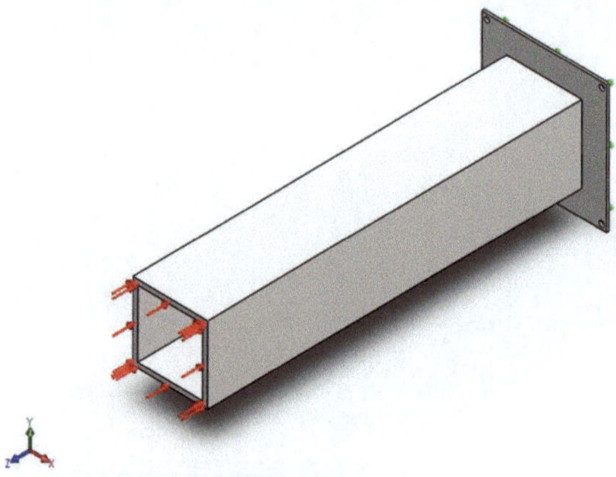

Fig. 5 Pylon template

3.4.2 Simulation

See Fig. 6.

3.5 Interpretation of Results

The maximum displacement of the guide is 4566×10^{-6} mm, it is a negligible value, so the guide is resistant to deformation, the maximum stress of Von Mises is 3.25×10^4 Pa, the yield strength Re = 1.75×10^8 Pa

$$\sigma ad = Re/s = 1.75 \times 10^8/1.5 = 1.16 \times 10^8\,Pa$$

- $\sigma VM < \sigma ad$: Therefore the resistance of the guide is validated.

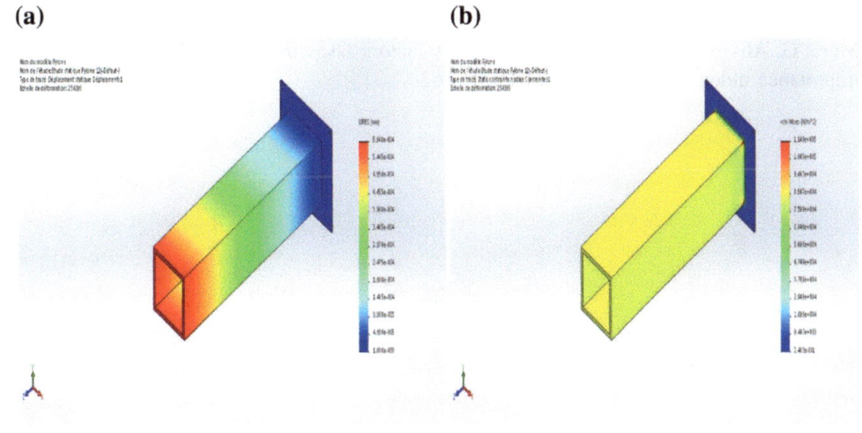

Fig. 6 **a** Moving the pylon, **b** Constraint of Von Misesa

Same methodology used the validation of the pylon. So the support works in the elastic domain and it resists the constraints imposed during operation.

4 Conclusion

In the present work, a design of a parabolic solar concentrator of opening diameter of 3 m. According to the reports given by "SOLIDWORKS Simulation", the maximum stress of Von Mises σVM in the most stressed zones for the mechanism remains lower than the admissible stress σad. Therefore, in static mode, the forces exerted by the wind and the weight do not pose any danger for their resistances. This solar concentrator can be used directly for several applications such as water heating, electricity generation using Stirling engine, vapour production, etc.…it will be interesting to study a solar concentrator with diameter aperture of 8.5 m with an application for Stirling system. It would be interesting to compare our geometry used form simulate pylon with other geometries as circular pylon for example.

References

1. Bernard J (2011) Energie solaire calculs et optimisation, Génie énergétique chapitre VI, la concentration solaire, pp 177–206
2. El Ouederni AR, Ben Salah M, Askri F, Ben Nasrallah M, Aloui F (2009) Experimental study of a parabolic solar concentrator. Revue des Energies Renouvelables 12(3):395–404
3. IRENA working paper (2012) Concentrating solar power renewable energy technologies: cost analysis series
4. Escobar-Romero JFM, Vázquez S, Montiel Y, Granados-Agustín F, Cruz-Martínez M, Rodríguez-Rivera E, Martínez-Yáñez L (2011) Building a parabolic solar concentrator prototype. J Phys Conf

5. Stine B, Geyer M (2001) Power from the sun. Lyle centre for regenerative studies
6. Murat G, Ahsan B (2018) Construction safety factors assessment through frequency adjusted importance index. Int J Ind Ergon 64:155–162

Sliding Wear Properties of Palm/Glass Fiber Hybrid Reinforced Vinylester Resin

Abderrazek Merzoug[1,2,3(✉)], Yasin Akgul[2,3], Bachir Bouhamida[1], Zouaoui Sereir[1], and Ali Kilic[2,3]

[1] Laboratoire de Structures de Composites et Matériauxinnovants, FacultédeGénieMécanique, Universitédes Sciences et de la Technologie d'Oran, USTO, BP 1505 El M'naouer, Oran of south Algeria, Algeria
merzoug_abdelrezak@outlook.fr
[2] TEMAG Labs, Faculty of Textile Technology and Design, Istanbul Technical University, Istanbul, Turkey
yasinakgul88@gmail.com
[3] Areka Advanced LLC, Istanbul, Turkey

Abstract. In this work, an experimental investigation was proposed for the evaluation the wear resistance of hybrid polymer composite inter-layer of Petiole date palm fiber (PDPF) and glass fiber (GF). In order to enhance the interfacial of hydrophobic fibers (PDPF) and hydrophilic (Vinylester resin), chemical treatment was proposed based on 5% NaOH for 48 h. Besides, the glass fiber (woven) was used in order to increase the specific stiffness and the dimensional stability of specimens. The manufacturing of four hybrids specimens was done using a VARTM process [(30PDPF/0GF), (20PDPF/10GF), (10PDPF/20GF) and (0PDPF/30GF)]. The wear characteristics were evaluated at a sliding distance of (1000 m), sliding velocity (1.7 m/s) and applied loads (20 and 40 N). Scanning electron microscopy was performed on both treated alkali fiber and worn surfaces of hybrid specimens to evaluate the surface modification and wear mechanisms, respectively. The adding of glass fiber to PDP fibers was advantageous on the wear resistance, where it improved the wear resistance of 30PDPF/0GF around 190 and 310% as compared to (20PDPF/10GF) and (10PDPF/20GF), respectively. The adding of loads was disproportional to wear resistance. The SEM of worn surfaces showed debonding and bending of fibers, and fragmentation, fiber broken, and deformation on the resinous regions. In the end, the hybridization by 20% of glass fibers exhibits a high wear resistance relative to its nature (partially bio-friendly) and price.

Keywords: Petiole date palm fibers · Hybridization · VARTM process · Wear resistance · Worn surfaces

1 Introduction

The polymer composite materials are covering wide interest in various engineering fields. In particular, the marine structure application, because these materials exhibit low density, high specific strength and specific modulus as compared to the

F. Chaari et al. (eds.), *Advances in Materials, Mechanics and Manufacturing*,
Lecture Notes in Mechanical Engineering,
https://doi.org/10.1007/978-3-030-24247-3_12

conventional materials (metal and alloys) [1]. Due to the easiest manufacturing, low cost and lightweight, the composite materials dominate in many other applications such as pipeline, high speed vehicles, water turbine and aircraft engine blades [2]. In contrast, the fibers used as reinforcement in these composite are petrochemical nature (Carbon, graphite and glass fibers … etc.). Where, it has negative effects on the environment and also involves various health hazards [3].

Thanks to the bio-friendly, low density, less high mechanical properties and as well as the economical benefit of industries, the natural fibers are becoming better alternatives to the synthetic fibers as reinforcements [4]. These natural fibers are composed of lingo-cellulose components, which contain strongly polarized hydroxyl groups [5]. The hydrophilic nature of these fibers leads to poor adhesion with hydrophobic matrices. Therefore, the combination of both constituents gives high water absorption. The interfacial adhesion is one of the most efficient parameters to obtain good and stable properties. This last can be improved by the modification of the matrix and/or natural fibers. In order to improve the surface modification of fibers, many studies have been done using chemical treatment [6–9]. The hybridization systems are proposed from many types of research to enhance the strength, stiffness and as well as moisture resistance and less cost ratio as compared to their properties. The hybrid composite can be composed of more than one kind of reinforcements in the same matrix, known as the combination of two reinforcements types be the most beneficial [10].

The presence of bio-composite materials in these application fields required high tribological properties. The industry and scientific researchers have met a high case number of failure mechanical parts when being subjected to tribological loading condition [11]. Therefore, it becomes important to evaluate the wear and frictional characteristics of bio-composite materials. Few studies have been done on the tribological performance of natural fiber reinforced polymeric matrixes. Shuhimi et al. [12] studied the tribological performances of epoxy resin reinforced by kenaf fiber and oil palm fiber under temperature effects and content amount; they obtained that increasing of temperature decreases the friction coefficient and increase the wear rates. They also showed that fiber content amount increases the wear resistance for both composites. Stapathy et al. [13] achieved an implementation of the artificial neural network (ANN) in analyzing the wear behavior of wood pine dust reinforced new class of epoxy filled matrix. They found that enhancement of sliding wear is by the adding pine wood dust. Besides, the filler content sliding velocity and normal load factors are affecting significantly on the specific wear rate. Yousif et al. [14], compared the friction coefficient and wear resistance of untreated and alkali treatment of oil palm as reinforcement of polyester resin. The sliding distance, the load applied and sliding velocity (0.85–3 km, 30–100 N and 1.7–3.9 m/s respectively) were the parameters studied. The untreated and treated oil-palm fibers improved the properties of the composite by about 40–70% and 40–80%. Besides, the interfacial between fibers and resin has highly influenced the characteristics.

Vinyl ester resin is a mixture of epoxy and polyester resins. Where, it gives good properties as compared to the easiest manufacturing and lower costs. The authors study the wear resistance properties of Petiole Date Palm/glass Fiber combined with Vinylester Resin composite material using pin-on test. To enhance the properties, the authors made an alkali treatment of PDPF fiber and added different amounts of glass fibers. Also, is an attempt to reduce the synthetic fibers amount in the environment and valorization the waste of biomass. Based on SEM observations, the interpretation of the worn surfaces was done.

2 Experimental

2.1 Materials

Date palm tree (Phoenix dactylifera) is a closely connected to the human life, it is grown in arid and semi-arid region of the world. Algeria has over than 18 million trees which occupy 1640 km^2. Every picking season, large quantities of date palm petiole, rachis and leave waster are accumulated in the agricultural land [3]. The fibers were obtained from a petiole date palm (PDP) tree part of Biskra region. The extraction of fiber method consists to immerse the petiole wood in a water bath for 72 h in room temperature. Where the wood will be soft, we can easily remove the fibers by hand. To enhance the surface modification of fibers, an alkali treatment of 5% NaOH for 48 h was done. In order to increase the strength and stiffness, the E-glass fibers were obtained in roving form from Dostkimya (Istanbul, Turkey). Its areal density was 49 g/m^2. Vinyl ester resin (Poliya 702), methyl ethyl ketone peroxide (MEK-P) and cobalt naphthenate 6% were obtained from Dost Kimya Tuzla, Istanbul, Turkey (Fig. 1).

2.1.1 Morphology Analysis

The PDPF diameter is measured using a ZEISS optical microscope equipped with a Moticam 2500 camera digitally controlled by Motic Images Plus V2.0 image processing program. Besides, Carl Zeiss ultra plus Gemini FESEM scanning electronic microscopy was used for the examination surface fiber modification before and after treatment. Where the magnification was used is 500X. The fibers were coated with gold before subjecting it to SEM. the surface structure of untreated and treated of PDP fibers are shown in Figs. 2 and 3. From the comparison, the untreated fiber had smooth surfaces, covered by layers of substances like a wax and impurities. However, the surfaces morphologies of PDPF treated were rough and significantly better as compared to untreated fiber. Which, it could be attributed to the removal of the hemicelluloses and lignin components and packing of cellulose chains. Several works have known the same remarks [15–17]. This morphological changing was affected directly on the fibers diameters Fig. 2. Rachid et al. [3] have known the same remarks (Table 1).

2.2 Hybrid Composite Processing

After the washing and the drying of PDP fibers, we chopped the length of PDPFs (60–120 mm) to prepare nonwoven mat of 600 g/m^2 areal density. Four inter-laminate composites based on PDPFs/glass woven fabric were prepared. Where, the weight fraction was fixed to 30:70 of the reinforcement and the matrix, respectively. The glass fabric was adopted on the top and the bottom of PDPFs. The configuration of specimens is given in Table 2.

Fig. 1 Preparation of the alkali treated PDPFs, **a** Date palm tree, **b** Petiole of date palm wood, **c** Extracted fibers, **d** drying in oven at 50 °C

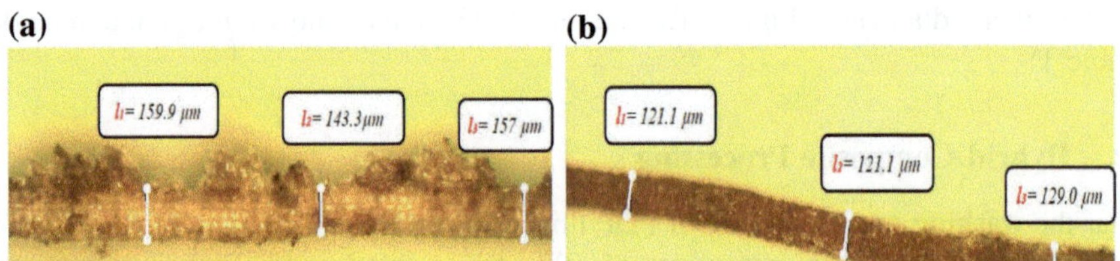

Fig. 2 Optical microscopy image of longitudinal surfaces of PDPF, **a** untreated fiber, **b** 5% NaOH for 48 h

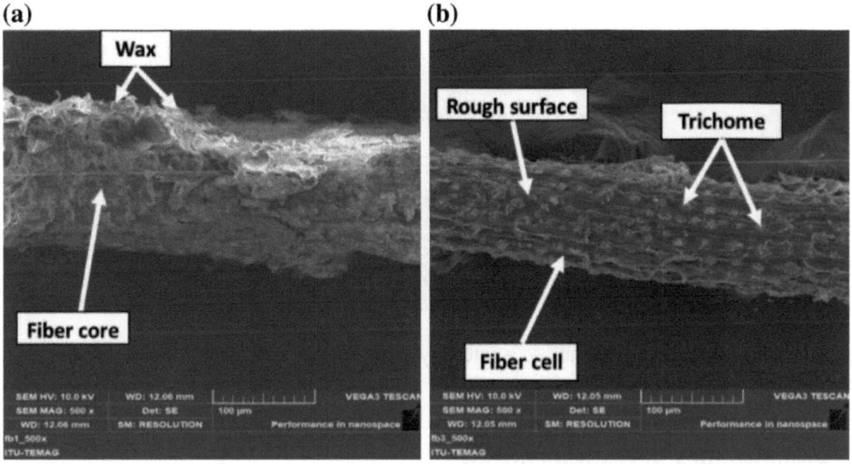

Fig. 3 Scanning electron microscopy (SEM) of longitudinal surfaces of PDPF, **a** untreated fiber, **b** 5% NaOH for 48 h

Table 1 Measurements of PDP fibers diameters

Fibers	Untreated	Treated
Diameters (μm)	145.54 ± 2.26	134.14 ± 3.34

Table 2 Hybrids specimens configuration

Hybrid specimens	PDP fiber (%)	Glass fiber (%)	Vinylester resin (%)
30PDPF/0GF	30	0	70
20PDPF/10GF	20	10	70
10PDPF/20GF	10	20	70
0PDPF/30GF	0	30	70

The vacuum resin transfer assistant (VARTM) process was used for the preparation of composites hybrid. All samples were fabricated over a cleaned glass plate. Polyurethane release film was placed on the top of the glass surface for easier removing of the final composite. To ensure the resin flow across the samples and established an even mold, two spiral pipes were then placed on the top edges of top layer for inlet and outlet resin, the first was connected to the resin bucket and the other connect to vacuum pump. After, peel ply and perforated film were added on the top surfaces of the hybrid composite, respectively. In order to prepare the resin mixture and injected inside, the infusion system was working well. Samples were set to consolidate for 24 h (Fig. 4).

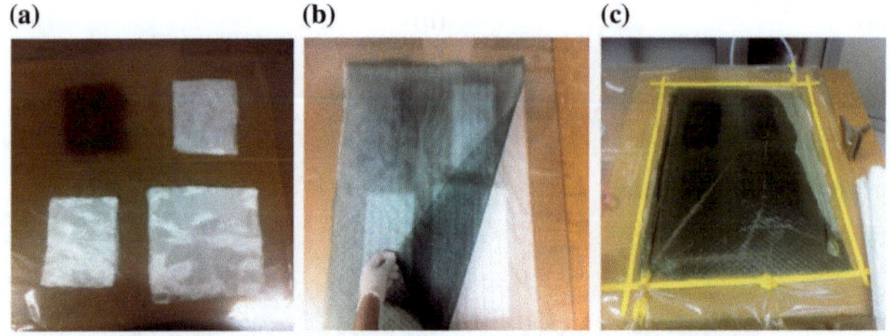

Fig. 4 VARTM process preparation, **a** inter-layer the samples, **b** adding the perforated release film, **c** lunch the process

3 Specimens Characterization

3.1 Wear Test

The wear resistance of the hybrids samples based on PDPF and GF were evaluated using a pin-on-disc tester. UTS Tribometer Test Machine (Karabuk University, Turkey) was used to investigate wear properties of samples at 20–25 °C temperature and the relative humidity of 30–45%. The wear tests were performed for all samples under loads of 20 and 40 N. Sliding distance and velocity were kept constant as 1000 m and 1.7 m/s, respectively (Fig. 5).

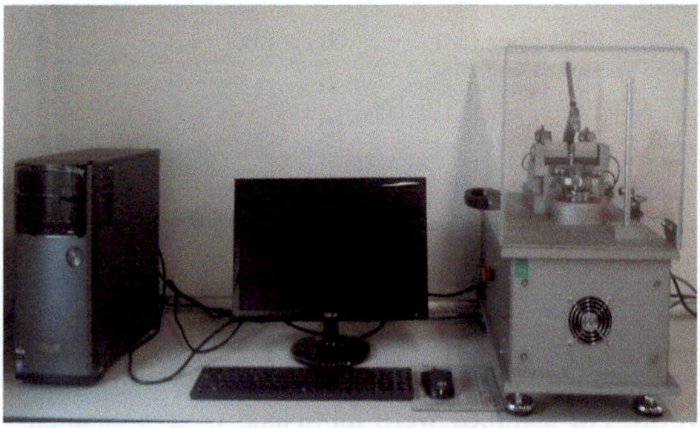

Fig. 5 UTS tribometer t10/20 device

3.2 Morphology Analysis

The worn surfaces of the hybrid specimens after testing were studied using Field Emission Scanning electron microscopy (Carl Zeiss Ultra Plus). Before taking the micrographs, the specimens were coated with a gold using plasma device. Due to their sensible state, magnifications ranging from 100X to 1 KX were proposed.

4 Results and Discussion

4.1 Wear Rate

The weight loss of the hybrid samples was determined after testing, using a balance of ±0.01 mg. following the relation (2 and 3), the wear rate (WR) and specific wear rate (SWR) was determined.

Wear rate:

$$WR = \frac{\Delta m}{\rho L}$$

The specific wear rate is the ratio of wear rate by the load applied *Fn*.

A Specific wear rate:

$$SWR = \frac{WR}{Fn} = \frac{\Delta m}{\rho L} \cdot \frac{1}{Fn} \quad (3)$$

where, Δm is the loss weight recorded during the test (g). ρ is the specimen density in (g/cm^3) and L is the sliding distance traveled (m).

Figure 6 illustrates the specific wear rate of hybrid specimens having different PDPF and glass fibers amount. The hybrid 0PDPF/30GF shows the highest wear resistance (0.015×10^{-2} mm^3/Nm), followed by 10PDPF/20GF. Due to its good interfacial with the matrix, high mechanical properties and its fiber direction (normal to the sliding movements) [18], the glass fabric has increased significantly the wear resistance of hybrid specimens. However, the lowest wear resistance is noted to 30PDPF/0GF hybrid specimens, due to the low mechanical properties of PDPF, the random distribution and its poor adhesion with vinyl ester resin comparatively to GF/resin.

The effect of applied load on specific wear rate for PDPF/GF hybrid reinforced vinylester is shown in Fig. 7. It observes an increase in specific wear rate with an increase in applied load for all samples. The specific wear rate of 0PDPF/30GF was increased by 64% as compared to that specific wear rate under 20 N. Besides, the 30PDPF/0GF was slightly increased by 34%. The proportionality between the specific wear rate and applied load implies a higher weight loss rate. The authors achieved similar results for polymer composite materials [19]. In comparison, when we have applied 40 N, we observed a distortion and instability dimensional of samples at different levels. It could be due to the repeated high axial trust while sliding. Besides, the poor tribological performance of polymer composite material and its low thermal property can affect on the dimensional stability. Therefore, the samples got stuck on counterface at higher load. Due to the direction of fibers and the combination of high thermal properties (GF) and the good viscoelastic properties (PDPF), the hybrid specimen 10PDPF/20GF has known the highest wear resistance.

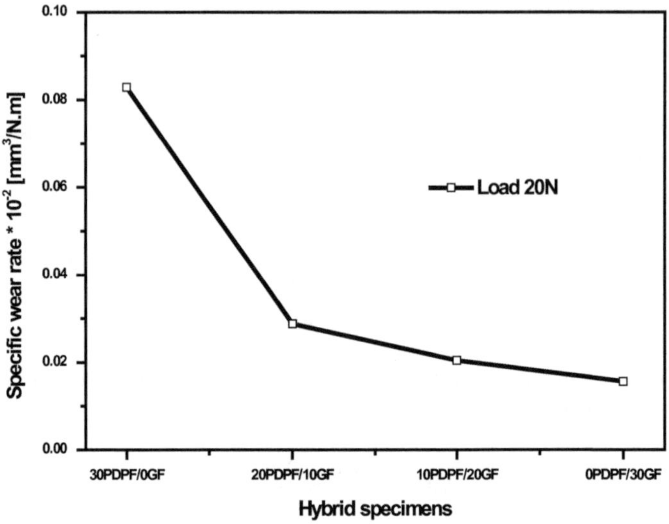

Fig. 6 Graph of specific wear rate versus hybrid specimens

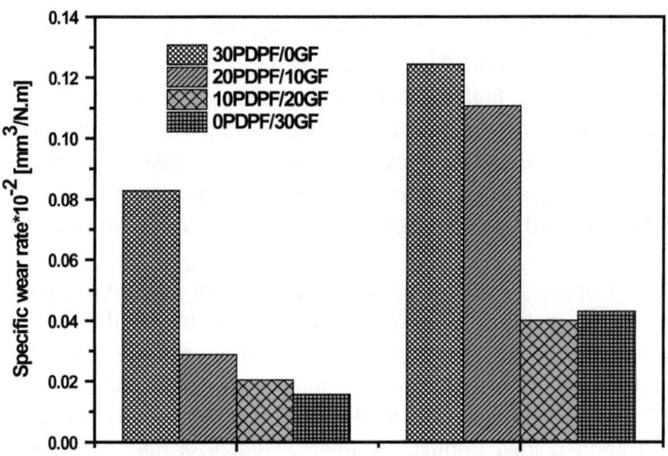

Fig. 7 Effects of loading on the specific wear rate of PDPF/GF hybrid specimens

4.2 Morphology Analysis

Figure 8a, b, c, d shows the microscopic observation of worn hybrid specimens under 20 N loading. Removal top layer of resin, fibers debonded, micro-pitting, fragmentation and partially broken of glass fibers were the main wear mechanisms. The Pits were produced due to adhesion between pin and counterface. Figure 8a shows the worn surface of pure PDPF reinforced the vinyl ester resin. Where, the effect of alkali treatment of fibers was considerably (good interfacial). Further, 30PDPF/0GF hybrid had known a removal of higher resin material comparatively to the others, due to the random distribution of fibers. Plus, it could be due to the presence of micro-porosity [14]. The adding of glass fibers is influenced significantly on the resin interfacial (b, c and d). The glass fabric layout showed an important role in wear resistance. Where, the direction of fiber in parallel to the normal force resisted more than that perpendicular direction (tensile stress resistance). However, the other direction was easy to be damaged and broken.

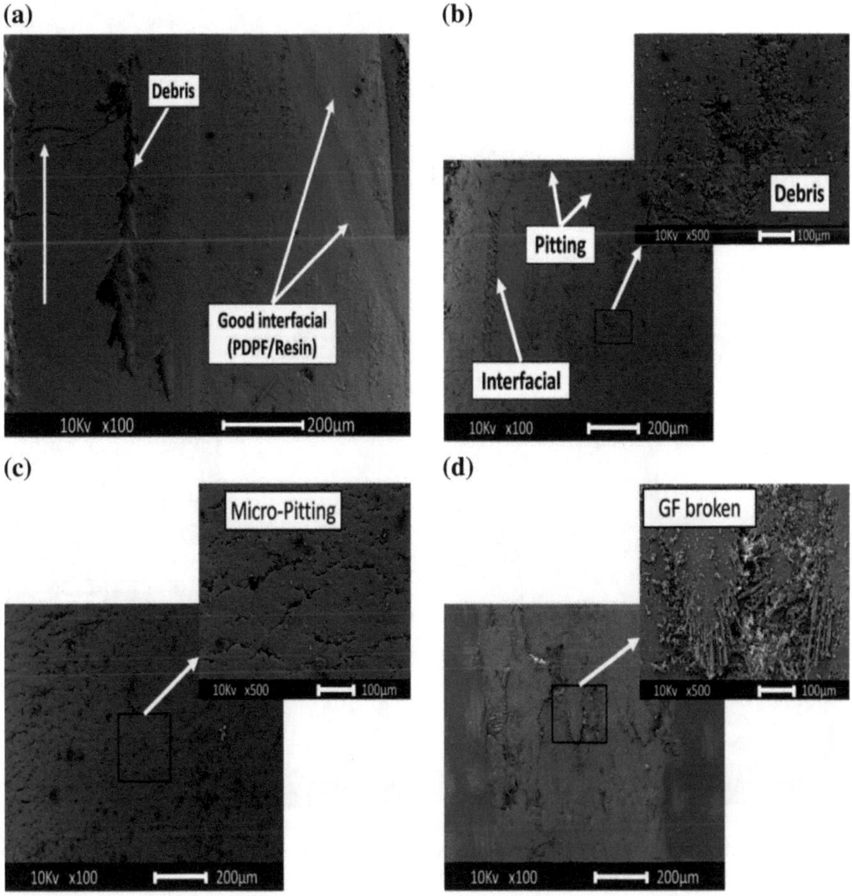

Fig. 8 SEM of worn surface of hybrid specimens under 20 N loading: **a** 30PDPF/0GF, **b** 20PDPF/10GF, **c** 10PDPF/0GF and **d** 00PDPF/30GF

Figure 9a, b, c, d shows the microscopic observation of worn hybrid specimens under 40 N loading. The increasing of loading was intense considerably to the wear mechanisms. Moreover, due to the high thermo-mechanical loading and high sliding distance (1000 m) leads to the softening of hybrid surfaces. Where, it produced fragmentations and broken fibers [14]. The hybrid 10PDPF/20GF shows the high dimensional stability surfaces as compared to others. It could be due to both properties (thermo-mechanical and viscoelastic) of glass fiber and petiole date palm fiber, respectively. We suggest that the use of PDPF like cores, because they are more appropriate to enhance the wear resistance of synthetic fibers and reducing the environmental impact.

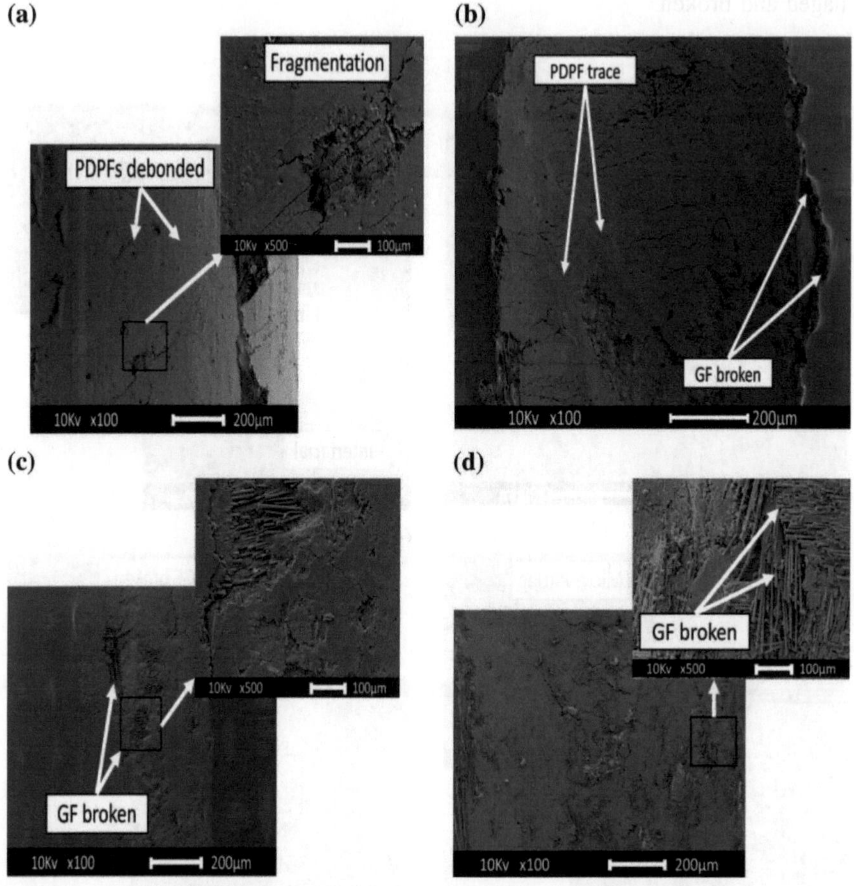

Fig. 9 SEM of worn surface of hybrid specimens under 40 N loading: **a** 30PDPF/0GF, **b** 20PDPF/10GF, **c** 10PDPF/0GF and **d** 00PDPF/30GF

5 Conclusion

The following points are drawn from the experimental investigation results:

1. The alkali treatment at 5% NaOH for 48 h was affected significantly on the surface modification of fibers. Where, it directly eliminated the impurities and wax layer around the fiber core. In parallel, the diameter was reduced.

2. The adding of glass fiber amount increased the wear resistance considerably of hybrid specimens by 190 and 300%. Because of its high thermo-mechanical properties of glass fibers and fiber orientation (roving woven).
3. The wear mechanism of hybrids specimens was predominated by removal and sliding of vinyl ester (especially in pure PDPF hybrid), debonding, micro-pitting, fragmentation and perpendicular (to the sliding direction) glass fiber broken.
4. The worn surface of 10PDPF/20GF shows less damage compared to others. There was the optimum amount combination of both viscoelastic (PDPF) and thermo-mechanical properties (GF).

Acknowledgements. The authors would like give a grateful acknowledges to AREKA LLC for the financial support of this work.

References

1. Xu H et al (2018) Static and dynamic bending behaviors of carbon fiber reinforced composite cantilever cylinders. Compos Struct 201(May):893–901
2. Raghavendra G, Ojha S, Acharya SK, Pal SK (2014) Jute fiber reinforced epoxy composites and comparison with the glass and neat epoxy composites. J Compos Mater 48(20):2537–2547
3. Benzidane R, Sereir Z, Bennegadi ML, Doumalin P, Poilâne C (2018) Morphology, static and fatigue behavior of a natural UD composite: the date palm petiole 'wood'. Compos Struct 203:110–123
4. Sankar PH, Reddy YVM, Reddy KH (2014) The effect of fiber length on tensile properties of polyester resin composites reinforced by the fibers of Sansevieriatrifasciata. 8:7–13
5. Liu L, Yu J, Cheng L, Qu W (2009) Mechanical properties of poly(butylene succinate) (PBS) biocomposites reinforced with surface modified jute fibre. Compos Part A Appl Sci Manuf 40(5):669–674
6. Lu T, Jiang M, Jiang Z, Hui D, Wang Z, Zhou Z (2013) Effect of surface modification of bamboo cellulose fibers on mechanical properties of cellulose/epoxy composites. Compos Part B Eng 51:28–34
7. Rajeshkumar G, Hariharan V (2015) Characterization of phoenix sp. natural fiber as potential reinforcement of polymer composites. J Ind Text 46(3):667–683
8. Amroune S, Bezazi A, Belaadi A, Zhu C, Scarpa F (2015) Tensile mechanical properties and surface chemical sensitivity of technical fibres from date palm fruit branches (Phoenix dactylifera L.). Compos. PART A 71:95–106
9. Senthamaraikannan P, Kathiresan M (2018) Characterization of raw and alkali treated new natural cellulosic fiber from Cocciniagrandis. L. Carbohydr Polym 186:332–343
10. Panthapulakkal S, Sain M (2006) Injection-molded short hemp fiber/glass fiber- reinforced polypropylene hybrid composites—Mechanical, water absorption and thermal properties. J Appl Polym Sci 103(4):2432–2441
11. El-Tayeb NSM, Yousif BF, Yap TC (2008) An investigation on worn surfaces of chopped glass fibre reinforced polyester through SEM observations. Tribol Int 41(5):331–340
12. Shuhimi FF, Bin Abdollah MF, Kalam MA, Hassan M, Mustafa A, Amiruddin H (2016) Tribological characteristics comparison for oil palm fibre/epoxy and kenaf fibre/epoxy composites under dry sliding conditions. Tribol Int 101:247–254

13. Kranthi G, Satapathy A (2010) Evaluation and prediction of wear response of pine wood dust filled epoxy composites using neural computation. Comput Mater Sci 49(3):609–614
14. Yousif BF, El-Tayeb NSM (2008) Adhesive wear performance of T-OPRP and UT-OPRP composites. Tribol Lett 32(3):199–208
15. Liu XY, Dai GC (2007) Surface modification and micromechanical properties of jute fiber mat reinforced polypropylene composites. Express Polym Lett 1(5):299–307
16. Aly-Hassan MS, Elbadry EA, Hamada H (2015) Corrigendum to 'agro-residues: surface treatment and characterization of date palm tree fiber as composite reinforcement'. J Compos 2015:1
17. Islam MS, Pickering KL, Foreman NJ (2010) Influence of alkali fiber treatment and fiber processing on the mechanical properties of hemp/epoxy composites. J Appl Polym Sci 119 (6):3696–3707
18. Dwivedi UK, Chand N (2009) Influence of fibre orientation on friction and sliding wear behavior of jute fibre reinforced polyester composite. Appl Compos Mater 16(2):93–100
19. Bajpai PK, Singh I, Madaan J (2013) Tribological behavior of natural fiber reinforced PLA composites. Wear 297(1–2):829–840

Screening of Factors Influencing Phosphate-Based Geopolymers Consolidation Time, Using Plackett-Burman Design

M. Zribi(✉), B. Samet, and S. Baklouti

Laboratoire de Chimie Industrielle, Ecole Nationale d'Ingénieurs de Sfax, Université de Sfax, BP.W3038 Sfax, Tunisie
{zribi.mar, baklouti.samir}@gmail.com, sametbasma@yahoo.fr

Abstract. In this paper, Plackett–Burman design was employed for screening factors influencing phosphate-based geopolymers consolidation time, through comparing the differences between each factor at two levels. This design screens 9 factors through 12 experiments. Considered factors are: Si/P molar ratio, curing temperature, heating time, calcination of used aluminosilicate precursor, mold's condition, aluminosilicate particle size, chemical composition of aluminosilicate precursor, pH and acidic solution. Results show that curing temperature, calcination of aluminosilicate precursor, mold's condition and aluminosilicate particle size factors are responsible for 97.79% of the consolidation time variation. The most significant effect is attributed to curing temperature, with a contribution of 73.46%. This factor presents a negative effect. Thus, its increase generates the decrease of consolidation time. This result can be explained by the endothermic character of the geopolymeric reaction. In fact, the increase of the curing temperature activates the reaction which accelerates the kinetics of consolidation. The impact of the other factors like Si/P molar ratio, the heating time, nature of aluminosilicate precursor, pH values and the acidic solution is very small and it can be neglected, in the studied conditions. These factors did not present a kinetics influence and they haven't a direct effect on the consolidation time.

Keywords: Screening · Phosphate-based geopolymers · Consolidation time · Plackett-Burman design · Curing temperature

1 Introduction

Phosphate-based geopolymers present a new variety of inorganic materials. They have found a wide range of applications such as construction material, dental cements, and radioactive waste stabilization [7]. These applications are insured thanks to several proprieties observed in this kind of geopolymers like high mechanical strength [10, 12, 17], excellent dielectric properties [9] and important chemical stability [8]. In addition to these proprieties, the consolidation time necessitates to be investigated, since it represents an interesting issue for industrial applications. Related literature shows that this propriety is not yet well controlled.

F. Chaari et al. (eds.), *Advances in Materials, Mechanics and Manufacturing*, Lecture Notes in Mechanical Engineering,
https://doi.org/10.1007/978-3-030-24247-3_13

The purpose of this work is to screen several factors that can influence phosphate-based geopolymers consolidation time. According to literature, several factors can be studied: the first one is the Si/P molar ratio (with Si = number of silicon moles derived from alumino silicate source and P = number of phosphorus moles derived from phosphoric acid). Several works have shown the dependence of phosphate based geopolymer on this factor [6, 13]. The second proposed factor is curing temperature. The significant effect of this factor on geopolymers proprieties and synthesis was proven in different works that studied similar varieties of materials like fly-ash-based geopolymers [15]. The third factor that can be studied is the heating time. The literature shows few works that proved the effect of this factor on alkali based geopolymers [11]. Added to these factors, the aluminosilicate particle size can be a considerable factor on this study. Many works proved the dependence of different materials proprieties on this factor [4, 5, 12]. According to preliminary observations, the condition of the mold during the heating (opened or closed) can present another factor to be studied in this work. Also, the acidic solution that can be prepared immediately or before 24 h of the reaction with the aluminosilicate precursor [17] can affect the consolidation time owing to the released energy during the mixing of acid with water. Another factor, which is considered in this work, is the chemical composition of aluminosilicate precursor. The used precursor can present a mineralogical composition rich in different oxides like Tunisian clays [13] or free of impurities [10] like pure kaolin. The calcination of the aluminosilicate precursor presents also an important factor to be studied. Popularly, the used aluminosilicate precursor should be calcinated to increase its reactivity [14]. Added to that, the pH of the starting mixture can be probably a significant factor at any chemical synthesis. For the synthesis of phosphate-based geopolymers, this factor is not yet studied.

The study of the consolidation time dependence on all these factors may be addressed following two methods: classical method and statistical method [2]. Classical method necessitates that only one factor is changed at each experiment to control its effect. Since a large number of factors are considered for this study, it is not feasible to independently study each factor. An efficient way to screen factors with the minimum of experiments is the use of experimental design. This second method remains an essential technique in chemometric science.

Experimental designs present several types namely we note: Bayesian experimental design, Block design, Box–Behnken design, Central composite design, Fractional factorial design, Optimal design, Plackett-Burman design, Taguchi methods……Each design is used according to the purpose of the study and the number of factors that are considered to be investigated in the considered work. Thus, the specific questions that the experiment is intended to answer must be clearly identified before carrying out the experiment. With the goal of this work which is screening, several experimental designs like those used for optimization or for studying factors dependence and relationship are eliminated. As far as that goes, the choice becomes limited on some experimental designs that are suitable for screening as fractional factorial design, Plackett–Burman design, Taguchi's Orthogonal Arrays… However, with an important number of factors (9 factors), one of the most adequate designs is Plackett-Burman design, thanks to its ability to identify active factors using as few experimental runs as possible. This design is often used in literature for similar objects [1, 2]. It is considered as a two-level multifactorial design [16]. Through comparing the differences between each factor at two levels, several factors can be screened out and other influencing one can be retained.

2 Materials and Methods

2.1 Materials and Preparation

Two different aluminosilicate precursors were used in this work: kaolin provided by Imerys France and natural Tunisian clay. The chemical composition of both precursors is illustrated in Table 1. According to X-ray fluorescence element analysis, kaolin present a mineralogical composition free of impurities, when natural Tunisian clay presents a mineralogical composition rich in different oxides like K_2O (6.07%) and Fe_2O_3 (8.72%).

Table 1 Chemical composition of the used aluminosilicate precursors

Type of aluminosilicate precursor	Kaolin	Natural Tunisian clay
Oxides	%masse	%masse
Al_2O_3	47	19.49
SiO_2	51.91	49.03
CuO	0.01	3.94
Fe_2O_3	0.56	8.72
K_2O	0.07	6.07
P_2O_5	0.05	0.09

Also, a commercial phosphoric acid H_3PO_4 85% provided by the Scharlau-chémie (SA) society was used. In accordance with each assembly, the corresponding aluminosilicate precursor was mixed with the corresponding phosphoric acid solution. The obtained mixture was kept at the mentioned curing temperature (25 or 85 °C) during the heating time (the time at which the sample was conserved in oven) chosen for the studied assembly (1 day or 7 days). The calcination of aluminosilicate precursors was carried out at 750 °C for 3 h in a static bed oven. The consolidation time of each sample was checked each 12 h. This time is determined according to the evolution of penetration depth in geopolymers pastes.

2.2 Plackett-Burman Design

The Plackett-Burman design allows to study k factors through N experiments ($N \geq k + 1$), where N must be a multiple of 4. The different factors considered in this work are summarized in Table 2, with their high and low levels. Since the number of factors used in this study was fixed at 9, a design consisting of 12 experiments ($N = 12$) was used, as presented in Table 3.

Table 2 Considered factors for the Plackett-Burman design and their high and low levels

Factors levels	Lower level (−1)	Higher level (+1)
Si/P molar ratio (X1)	0.5	2
Curing temperature (X2)	25 °C	85 °C
Heating time (X3)	1 day	7 days
aluminosilicate particle size (X4)	≥ 63 μm	≤ 63 μm
Mold's condition (X5)	Opened	Closed
Acidic solution (X6)	A[a]	B[b]
Chemical composition of aluminosilicate precursor (X7)	Kaolin	Natural Tunisian clay
Calcination of aluminosilicate precursor (X8)	With	Without
pH (X9)	Low (between 1 and 3)	High (between 4 and 5)

[a]Prepared immediately
[b]Prepared 24 h before mixing with aluminosilicate precursor

In this design, each experiment contained only either the +1 or −1 value of each variable. However, it presents a completely different variable represented by +1 or −1 compared to another experiment. According to the Plackett-Burman design rules, none of the twelve experiments was similar to the other [1]. For each experiment, the measured response is recorded as illustrated in Table 4.

3 Results and Discussions

The influence of all factors on the phosphate-based geopolymers consolidation time is presented in Fig. 1. Figure 1a presents the bar graph of factors effects. The surface of each bar is proportional to the importance of the considering factor impact. Figure 1b, c represent the individual and cumulative Pareto effects, respectively.

The design results show that the most significant effect corresponds to curing temperature factor. In fact, according to the Pareto diagram illustrated in Fig. 1b, this factor participates by 73.46% in the sensitivity of the consolidation time. Also, curing temperature presents a negative effect which means that its increase generates the decrease of consolidation time. This result can be explained by the endothermic character of the geopolymeric reaction. In fact, the increase of the curing temperature activates the reaction which accelerates the kinetics of consolidation. Added to this dominant factor, calcinations, mold's condition and particle size factors have a modest effect on the discussed response. As shown in Fig. 1b, the effects of these factors can be classified as following: X8 > X5 > X4. The calcination factor participates by 15.01% in the variation of the consolidation time, when mold's condition and particle size participate by 5.33 and 3.98%, respectively.

According to the sign of contribution illustrated in Fig. 1a, we can conclude that: The calcination of precursors participates in the decreasing of the consolidation time, thanks to its influence on the reactivity of the starting material. The calcination

Table 3 Experimental design

No. exp.	Si/P molar ratio	Curing temperature	Time of heating	Particles size	Mold's condition	Acidic solution	Nature of aluminosilicate	Calcination	pH
1	2	85	1 day	≤ 63	Closed	B	Kaolin	With	Low
2	0.5	85	7 days	≥ 63	Closed	B	Natural Tunisian clay	With	Low
3	2	25	7 days	≤ 63	Opened	B	Natural Tunisian clay	Without	Low
4	0.5	85	1 day	≤ 63	Closed	A	Natural Tunisian clay	Without	High
5	0.5	25	7 days	≥ 63	Closed	B	Kaolin	Without	High
6	0.5	25	1 day	≤ 63	Opened	B	Natural Tunisian clay	With	High
7	2	25	1 day	≥ 63	Closed	A	Natural Tunisian clay	Without	Low
8	2	85	1 day	≥ 63	Opened	B	Kaolin	Without	High
9	2	85	7 days	≥ 63	Opened	A	Natural Tunisian clay	With	High
10	0.5	85	7 days	≤ 63	Opened	A	Kaolin	Without	Low
11	2	25	7 days	≤ 63	Closed	A	Kaolin	With	High
12	0.5	25	1 day	≥ 63	Opened	A	Kaolin	With	Low

Table 4 Design matrix and results of Plackett-Burman

No. exp.	X1	X2	X3	X4	X5	X6	X7	X8	X9	Time of consolidation (h)
1	1	1	−1	1	1	1	−1	−1	−1	24
2	−1	1	1	−1	1	1	1	−1	−1	24
3	1	−1	1	1	−1	1	1	1	−1	96
4	−1	1	−1	1	1	−1	1	1	1	48
5	−1	−1	1	−1	1	1	−1	1	1	168
6	−1	−1	−1	1	−1	1	1	−1	1	72
7	1	−1	−1	−1	1	−1	1	1	−1	144
8	1	1	−1	−1	−1	1	−1	1	1	48
9	1	1	1	−1	−1	−1	1	−1	1	12
10	−1	1	1	1	−1	−1	−1	1	−1	24
11	1	−1	1	1	1	−1	−1	−1	1	96
12	−1	−1	−1	−1	−1	−1	−1	−1	−1	4

increases the reactivity of the aluminosilicate precursor [3]. As a result, the precursor becomes more able to react with the phosphoric acid and it forms the new solidified material faster.

The mold's condition presents a minor effect (5.33%) but it can be explained by the easy evaporation of the liquid excess when the mold is open. Consequently, the consolidation of material can be accelerated and the studied response decreased.

Finally, the particle size factor has the lowest impact on the consolidation time, that is equal to 3.98%. When the particle size decreases, the specific surface increases which ameliorates the reactivity of the precursor. As a result, the reaction becomes faster and the time of consolidation decreases. In totality, curing temperature, calcination, mold's condition and particle size factors explain more than 97.79% of the consolidation time variation, as presented in Fig. 1c. Consequently, to decrease the maximum of the consolidation time, we are recommended to use the highest possible curing temperature, an open mold and a calcined aluminosilicate precursor with the smallest particle size.

The impact of the other factors like Si/P molar ratio, the heating time, nature of aluminosilicate precursor, pH values and the acidic solution is very small and it can be neglected, in the studied conditions. These factors did not present a kinetics influence and they haven't a direct effect on the consolidation time.

4 Conclusion

A screening of different factors influencing phosphate-based geopolymers consolidation time is carried out in this work, using Plackett-Burman design. Nine factors were considered, according to literature. The impudence of each factor on the variation of the measured response is discussed. As a result, curing temperature, calcination, mold's condition and particle size factors are considered as the main responsible to explain

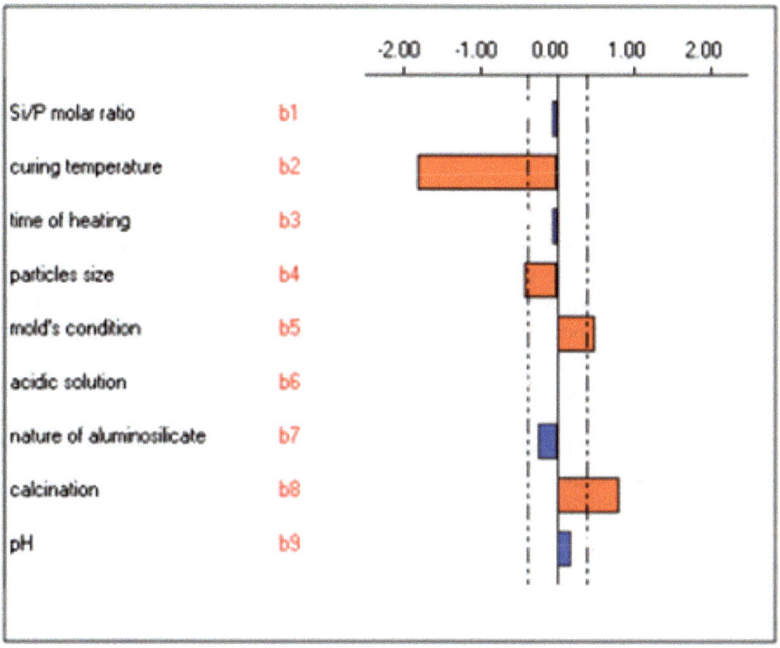

(a) Bar graph of factors effects

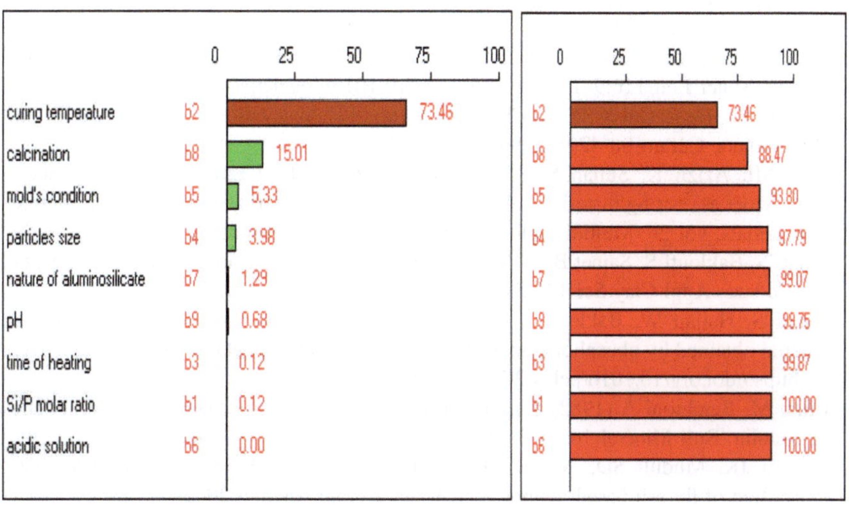

(b) Individual Pareto effects (c) Cumulative Pareto effects

Fig. 1 Screening results of the main factors of Plackett-Burman design more than 97.79% of the consolidation time variation. Added to that, curing temperature was selected as the most influencing factor with 70% of the consolidation time variation.

References

1. Amara AA, Salem-Bekhit Mounir M, Alanazi FK (2013) Plackett-Burman randomization method for bacterial ghosts preparation form E. coli JM109. Saudi Pharm J 6(2):1–7
2. Ahuja SK, Ferreira GM, Moreira AR (2004) Application of Plackett-Burman design and response surface methodology to achieve exponential growth for aggregated shipworm bacterium. Biotechnol Bioeng 85(6):666–675
3. Bich C (2005) Contribution à l'étude de l'activation thermique du kaolin: évolution de la structure cristallographique et activité pouzzolanique. Doctoral dissertation, Lyon, INSA
4. Bui DD, Hu J, Stroeven P (2005) Particle size effect on the strength of rice husk ash blended gap-graded portland cement concrete. Cement Concr Compos 27(3):357–366
5. Buffat P, Borel JP (1976) Size effect on the melting temperature of gold particles. Phys Rev A 13(6):2287
6. Cao D, Su D, Lu B, Yang Y (2005) synthesis and structure characterization of geopolymeric material based on metakaolinite and phosphoric acid. Chin Ceram J 33
7. Davidovits J (2008) Geopolymer chemistry and applications, 2nd edn. Inst. Geopolymer, Sain Quentin, France
8. Davidovits J (2011) Geopolymers-chemistry and applications, 4th edn. Institut géopolymère, Saint-Quentin
9. Douiri H, Louati S, Baklouti S, Arous M, Fakhfakh Z (2014) Structural, thermal and dielectric properties of phosphoric acid-based geopolymers with different amounts of H3PO4. Mater Lett 116:9–12. https://doi.org/10.1016/j.matlet.2013.10.075
10. He Y, Liu L, He L, Cui X (2016) Characterization of chemosynthetic H3PO4–Al2O3–2SiO2 geopolymers. Ceram Int 42(9):10908–10912
11. Khan MI, Azizli K, Sufian S, Siyal AA, Man Z, Ullah H (2014) Sodium silicate free geopolymer as coating material: adhesion to steel. In: 1st International electronic conference on materials, vol 26. Multidisciplinary Digital Publishing Institute
12. Louati S, Baklouti S, Samet B (2016) Acid based geopolymerization kinetics: effect of clay particle size. Appl Clay Sci 132:571–578
13. Louati S, Hajjaji W, Baklouti S, Samet B (2014) Structure and properties of new eco material obtained by phosphoric acid attack of natural Tunisian clay. Appl Clay Sci 101:60–67. https://doi.org/10.1016/j.clay.2014.07.015
14. Murat M, Barchioni A (1982) Corrélation entre l'état d'amorphisation et l'hydraulicité du métakaolin. Bull Minéral 105:543–555
15. Sadangi JK, Muduli SD, Nayak BD, Mishra BK (2013) Effect of phosphate ions on preparation of fly ash based geopolymer. IOSR J Appl Chem 4:20–26
16. Stanbury PF, Whitaker A, Hall SJ (1986) Media for industrial fermentations. Principles of fermentation technology. Pergamon, Oxford, pp 93–122
17. Tchakouté HK, Rüscher CH, Kamseu E, Andreola F, Leonelli C (2017) Influence of the molar concentration of phosphoric acid solution on the properties of metakaolin-phosphate-based geopolymer cements. Appl Clay Sci 147:184–194. https://doi.org/10.1016/j.clay.2017.07.036

Effect of Droplet Initial Temperature on Substrate Melting and Its Re-solidification in Plasma Spray Process

M. Driouche[1(✉)], T. Rezoug[1], and M. El-Ganaoui[2]

[1] Laboratoire des sciences aéronautiques, Institut d'aéronautique et des études spatiales, Université Blida1, Blida, Algérie
drioucheaero@yahoo.fr

[2] Laboratoire LERMAB, Université de la lorraine, Metz, France

Abstract. The properties of the coatings produced by plasma spraying are essentially related to the adhesion force between the projected particles and the substrate. The melting of the substrate can significantly improve this adhesion strength. In this paper, a 2D axisymmetric model was developed to simulate the impact of a fully melted particle of alumina in droplet form on an aluminum substrate, taking into consideration the melting of the substrate and its re-solidification using Ansys Fluent 14. Equations of fluid dynamics and energy including phase change are solved in a structured mesh by finite volume discretization. The volume of fluid method (VOF) is used for tracking the free surface of the droplet. An enthalpy-porosity formulation is used to model the solidification. The effect of the initial temperature of the droplet on the melting of the substrate has been studied. Substrate fusion was observed before the droplet spread completely. The volume of molten substrate increases over time, to reach its maximum, then it begins to decrease because of its re-solidification. The substrate is completely solidified well after the solidification of the droplet. It has been observed that the volume of the molten substrate is greater when the initial temperature of the droplet is important, which can improve the adhesion strength of the coating.

Keywords: Droplet impact · Substrate melting · Solidification · CFD · Plasma spray

1 Introduction

Plasma spray process is used in industry to form coatings resistant to corrosion, oxidation, wear, etc. The coating is formed by successive stacking of lamellae, consisting of melted particles in the form of a droplet, and accelerated by the plasma jet, which spread and solidify on a solid substrate. The properties of the coating thus produced depend essentially on the spread of the molten droplet on the substrate, its cooling, solidification, and the contact of the lamellae with the substrate and with each other. It should be noted that industrial applications generally require good adhesion to the substrate and good cohesion (adhesion between the deposited lamellae) [1–3]. The melted droplet contains enough heat energy, which can melt the substrate locally after

F. Chaari et al. (eds.), *Advances in Materials, Mechanics and Manufacturing*,
Lecture Notes in Mechanical Engineering,
https://doi.org/10.1007/978-3-030-24247-3_14

the impact. If substrate melting is important for good adhesion strength, it can also influence the splat morphology (splashing) [4].

Several works have been devoted to the understanding of the process of the formation of the coating, by experimental and numerical methods [1]. The time of spreading and solidification of the droplets takes only a few microseconds, and with the limits of the instruments, the experiment becomes very difficult [5]. Recently considerable effort has been directed towards numerical studies of the impact, spreading and solidification of the droplet [6–8]. Zhang [9] proposed a theoretical model to study the spreading and solidification of droplet. Surface tension, solidification and thermal contact resistance were included in this model. Bussmann et al. [10] have developed a 3D model of the flow of a free-surface fluid, which makes it possible to model the phenomenon of splashing induced by the spreading of the droplet, without taking into account the heat transfer between the droplet and the substrate. Pasandideh-Fard et al. [11] extended the Bussmann et al. [10] model taking into account heat transfer and solidification during spreading of the droplet.

The melting of the substrate has not been considered in the literature mentioned above. However, the experiments have noted the melting of the substrate with the impact of melted droplets [4, 12]. Wang et al. [13] numerically studied the cooling and solidification of splat and substrate, and deduced the maximum depth of melting of the substrate for various process parameters. Zarzalejo et al. [14] developed a numerical model to study the impact of a metallic droplet on a solid substrate, the effect of convection induced by the spreading of the droplet is modeled by an effective thermal conductivity, and the model provides the position of the melting front and the re-solidification in the substrate. Zhang et al. [15] experimentally and numerically studied splat morphology and substrate melting, following the impact of a molybdenum molten droplet on a molybdenum, mild steel and glass substrate. The effect of substrate nature and temperature on splat morphology was studied. Li et al. [16] developed a numerical model to examine the effect of droplet solidification, and the melting and re-solidification of the substrate by thermal spray. Experimental and numerical studies were performed for different combinations of droplets and substrates. It has been experimentally validated that the properties of the droplet and the substrate have a significant effect on the melting of the substrate. Rajesh et al. [17] have numerically studied the impact, spreading and solidification of a melted droplet on a substrate of different material taking into consideration the melting and re-solidification of the substrate. They observed that the fusion of the substrate for the studied case begins shortly after the droplet's impact. Most of the previous work generally considers the melting of substrate after the impact of millimeter metal droplet with low velocity, which is far from plasma spraying conditions. Very few numerical studies have considered the melting and re-solidification of the substrate during the impact and spreading of a melted ceramic droplet on a metallic substrate in a plasma spray conditions. In this paper, a 2D axisymmetric model to study the spreading and solidification of a droplet of alumina impacting an aluminum substrate, using the CFD method was developed, using Ansys Fluent 14. The melting and re-solidification of the substrate is taken into consideration in this model. The effect of the initial temperature of the droplet on the melting of the substrate is studied.

2 Numerical Model

Figure 1 represents the problem studied, a droplet of alumina 40 μm in diameter, impacted with velocity of 150 m/s, on an aluminum substrate initially at 450 K. The initial temperatures of the droplet at impact are: 2600, 2800 and 3000 K. In this model, transient fluid flow during impact, spreading, heat transfer, and solidification of the droplet are considered. The volume of fluid method (VOF) for tracking the free surface of the droplet is used. The evolution of the melting volume in the substrate is modeled by solving the equations of energy and solidification in the substrate. In this study the mixing between the droplet and the melted part of the substrate is not considered. The interface between the substrate and the droplet does not change and remains in its initial state. In addition, substrate melt ejection does not occur [17].

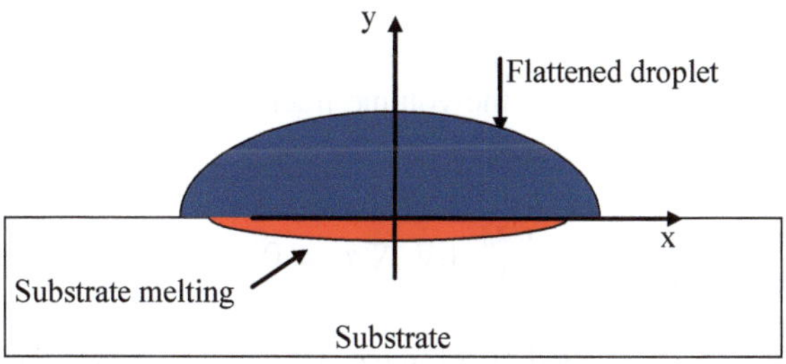

Fig. 1 Substrate smelting during spreading of the droplet

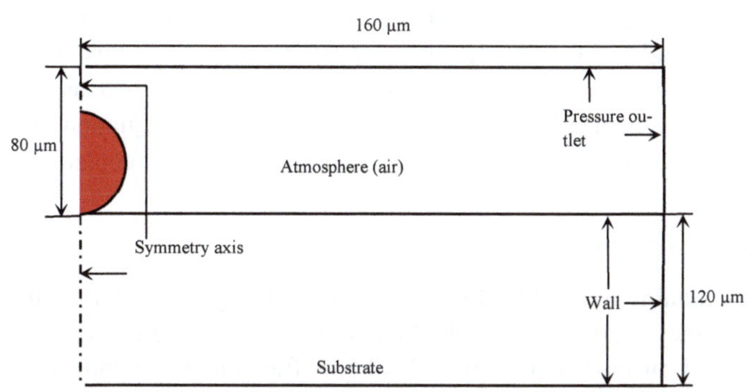

Fig. 2 2D axisymmetric domain for the simulation of the impact of the droplet on the substrate

2.1 The Fluid Dynamics

In this study, fluid dynamics of impact and spreading of the droplet is modeled using the finite volume method. The equations of the dynamics are solved in a 2D axisymmetric domain. The flow is considered incompressible, laminar and Newtonian; the conservation equations for mass and momentum are as follows:

$$\nabla \cdot \vec{v} = 0 \tag{1}$$

$$\frac{\partial(\rho\vec{v})}{\partial t} + \nabla \cdot (\rho\vec{v}\vec{v}) = -\nabla P + \nabla \cdot \left[\mu\left(\nabla\vec{v} + \nabla\vec{v}^T\right)\right] + \rho\vec{g} + F_{vol} + S_y \tag{2}$$

In these equations, $\vec{v}$ is the velocity vector, P the pressure, $\vec{g}$ gravitational acceleration, F_{vol} the surface tension force per unit volume, $[\mu(\nabla\vec{v} + \nabla\vec{v}^T)]$. The viscous stress vector, S_y the source term, ρ the density and μ is the dynamic viscosity. The volume of fluid meth (VOF) was used to capture the interface between two immiscible fluids. In this method a scalar function α_q is defined, which represents a fraction of the volume of a cell containing the qth fluid. Variables and properties in any cell are purely representative of a phase or a mixture of several phases. When the cell is empty of qth fluid, α_q equals to zero. If α_q is between zero and one, the cell contains the interface between the qth fluid and one or more other fluids, and if the cell is full of qth fluid α_q equals to one. The tracking of the moving interface between the phases is accomplished by solving the continuity equation for the volume fraction of one or more phases, for the qth phase, the equation is as follows:

$$\frac{\partial\alpha_q}{\partial t} + \vec{v} \cdot \nabla\alpha_q = 0 \tag{3}$$

The surface tension is modeled by a force of volume (F_{vol}) acting on the liquid at its free surface, which appears as a source term in Eq. (1). This force is calculated using the continuum surface force (CSF) model, developed by Brackbill et al. [18]. This model is also used to predict the curvature of the surface of the droplet near the wall using the computed tangential and normal flow velocity components along the wall, and the contact angle of the droplet at the substrate surface. In this study a constant contact angle of 90° is considered. A porosity function, known as Darcy law, is applied, which acts as a source term (S_y) in the momentum equation. The solidified liquid decelerates until the droplet becomes stationary due to complete solidification.

2.2 Heat Transfer and Solidification

The porosity-enthalpy method [19] is used to model the process of solidification of the droplet and the melting of the substrate. In this method the solidification interface is not explicitly tracked. Instead, a quantity called liquid fraction, which indicates the fraction of volume that is in a liquid state, is associated with each cell of the domain. An enthalpy balance is used in order to calculate the liquid fraction. The mushy zone is a region in which the liquid fraction is between zero and one. The total enthalpy in the droplet or substrate is introduced as the sum of sensible heat and latent heat as follows:

$$H = h + \Delta H \tag{4}$$

where

$$h = h_{ref} + \int_{T_{ref}}^{T} C_p dT \quad (5)$$

where h_{ref} and T_{ref} are respectively the sensible heat and the reference temperature, and C_p is the specific heat at constant pressure.

The liquid fraction β is defined as:

$$\begin{aligned} &\beta = 0 && if\ T < T_{solidus} \\ &\beta = \frac{T - T_{solidus}}{T_{liquidus} - T_{solidus}} && if\ T_{solidus} < T < T_{liquidus} \\ &\beta = 1 && if\ T > T_{liquidus} \end{aligned} \quad (6)$$

By the definition of the liquid fraction, the term of the latent heat becomes:

$$\Delta H = \beta L \quad (7)$$

where L is the latent heat of the material and it can vary between zero for the solid and L for the liquid. The equation of energy for solidification/melting problems can be written as:

$$\frac{\partial}{\partial t}(\rho H) + \nabla \cdot (\rho \vec{v} H) = \nabla \cdot (k \nabla T) \quad (8)$$

In which H should be substituted from Eq. 3, k is the thermal conductivity.

3 Results and Discussion

In order to validate the model presented in this article, the results obtained by the simulation are compared with those obtained experimentally. The results of the simulation of the impact of a tin droplet of 2.7 mm diameter on a stainless steel substrate under the same conditions of the experiments, made by Aziz and Chandra [20], are compared with those measured. Figure 3 shows the temporal evolution of the spreading. A spreading phase is observed up to the time 3.3 ms where the maximum of spread factor ξ (ξ is the ratio between the diameter D of the spreading droplet and its initial diameter D_0) reached is 1.96 against 2.05 measured by Aziz and Chandra [20]. Figure 3 shows a very good agreement between the values of the spread factor obtained by the simulation and those measured.

After the validation of the model presented, simulations were carried out, for the case of impact of alumina droplet on an aluminum substrate under the plasma spray conditions. A 2D-axisymmetric computational domain is considered (Fig. 2); the diameter and the initial velocity of the droplet are 40 μm and 150 m/s respectively. To study the effect of the initial temperature of the droplet on the melting of the substrate, three values are considered: 2600, 2800 and 3000 K. To take into account the non-perfect contact between the droplet and the substrate, a thermal contact resistance of

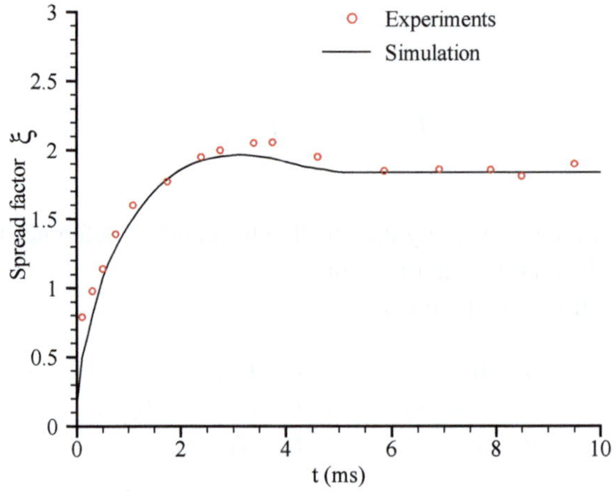

Fig. 3 Evolution of the calculated spread factor in comparison with experimental results [20], for the impact of a 2.7 mm diameter tin droplet

10^{-8} m^2 K/W is retained. The substrate is initially at 450 K, the other walls of the substrate are maintained at 300 K. The Thermophysical properties of the droplet and the substrate used in the simulations are given in Table 1.

Table 1 Thermophysical property data ([21, 22])

Properties	Alumina (Al_2O_3)	Aluminum (Al)
Density (kg/m^3)	3990	2390
Thermal conductivity (W/m K) (sol)	5.9	210
Thermal conductivity (W/m K) (liq)	7.86	105
Specific heat (J/kg K) (sol)	1273	1080
Specific heat (J/kg K) (liq)	1358	1080
Kinematic viscosity (m^2/s)	1.026×10^{-5}	
Surface tension (N/m)	0.69	
Melting Point (K)	2327	934
Latent heat of fusion (J/kg)	1.16×10^6	3.97×10^5

The left part of each picture in Fig. 4 shows the temporal evolution of the spreading of the droplet at 3000 K as well as its solidification, and the melting of the substrate and its re-solidification, represented by the liquid and solid fraction in the droplet and the substrate. The right side represents the temperature field. When the droplet impacts on the substrate, it starts to extend outwards. Due to the heat transferred from the droplet to the substrate, a thin layer of solid appears in the droplet at 0.2 μs. At this time the temperature of the substrate at the interface reaches its melting point temperature, however the melting of the substrate begins. After about 0.82 μs the spreading of the droplet is at its maximum to reach a diameter of 132 μm, which corresponds to spread factor of 3.3. The solidification of the droplet and the melting of the substrate are progressing. At 3.62 μs, the melting of the substrate reaches its maximum, while the droplet is not completely solidified, but the central part which is in contact with the melted portion of the substrate is solidified. This situation promotes a good adhesion between the coating and the substrate. After that, the substrate begins to solidify, due to the heat transfer to the depth of the substrate by conduction. The spread droplet is completely solidified while part of the substrate is still melted as shown in the figure at 6.53 μs. The substrate is re-solidified completely to about 8 μs. After spreading and solidification, the droplet is like a flattened disk with slightly raised rims, related to the accumulation of mass under the effect of surface tension forces.

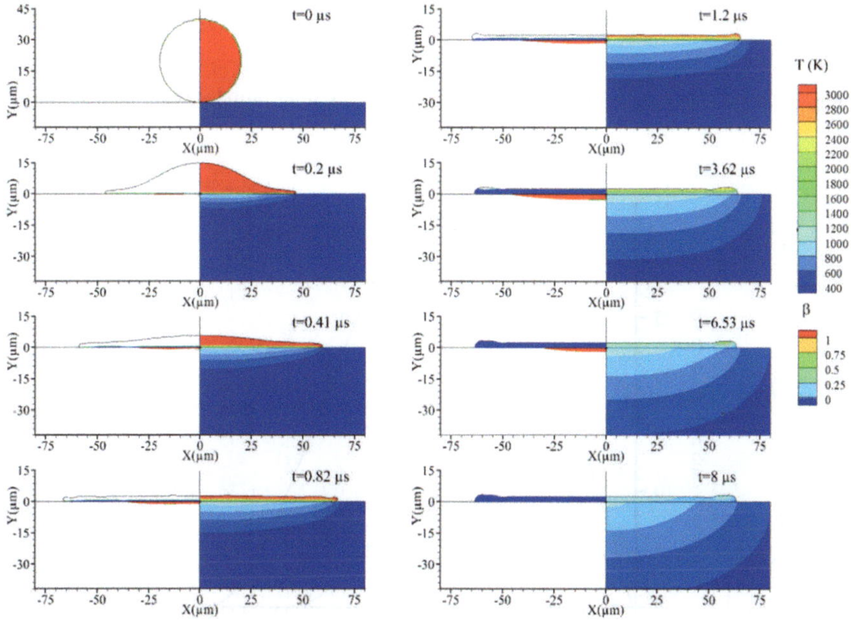

Fig. 4 Evolution of the spreading, the temperature of the alumina droplet at 3000 K and the melting and re-solidification of the aluminum substrate

The temperature history of the surface of the substrate for different initial droplet temperatures is shown in Fig. 5. It can be noted that the melting of the substrate is earlier when the initial temperature is important. The maximum temperature reached in the substrate is also proportional to this initial temperature. The cooling and re-solidification time is also greater for a droplet with a very high initial temperature.

Figure 6 illustrates the temporal evolution of the volume of molten substrate for different initial temperatures of the droplet. It can be observed that the melting of the substrate begins shortly after the droplet has been impacted with the substrate. For an initial temperature of 2600 K, the beginning of the melting of the substrate is at 0.4 μs,

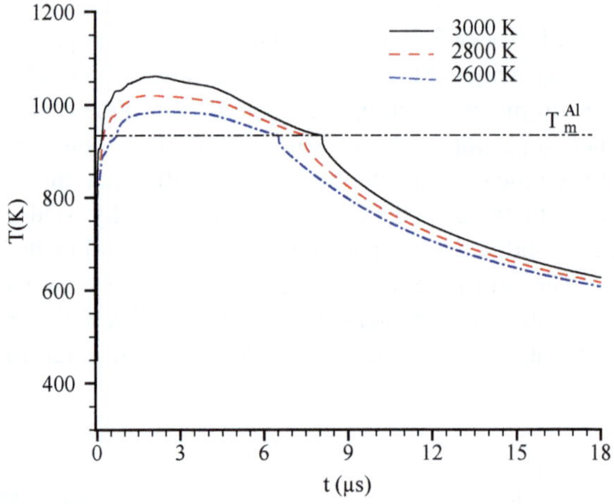

Fig. 5 Surface temperature of substrate history for different initial droplet temperatures

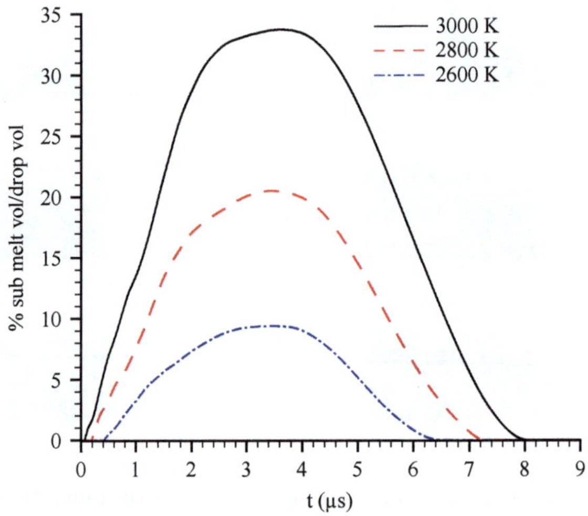

Fig. 6 Temporal evolution of the volume of the molten substrate for different initial temperatures of the droplet

and the end of re-solidification is at 6.5 μs, to reach a maximum volume of 9.38% of the volume of the droplet. The fusion of the substrate starts at 0.2 μs for an initial temperature of 2800 K, and its re-solidification is finished at 7.47 μs, the maximum volume reached is 20.84%. For an initial temperature of 3000 K, the maximum volume is 33.74% between the beginning of the fusion at 0.1 μs and the end of re-solidification at 8 μs. It can conclude that the volume of the molten substrate is proportional with the initial temperature of the droplet. This phenomenon is quite easy to understand because the droplet contains more energy at a higher temperature.

4 Conclusions

A 2D-axisymmetric model based on the finite volume method was developed to simulate the impact, the spreading, the solidification of a droplet of alumina on an aluminum substrate, as well as the melting of the substance and its re-solidification. The model used considers coupled phenomena, such as free surface movement, fluid dynamics, heat transfer and phase change (solidification/melting). The effects of thermal contact resistance and surface tension are taken into account in the model. The initial temperature of the droplet was varied to study its effect on melting and re-solidification of the substrate. The results are presented for the spreading and solidification of the droplet, as well as the melting and re-solidification of the substrate. These results are obtained by simulations in the plasma projection conditions. It was observed that the melting of the substrate begins shortly after the impact of the droplet. The melting of the substrate reached its maximum, while the droplet is close to its total solidification. A droplet initially at a high temperature contains more energy to further melt the substrate, which can promote good adhesion between the coating and the substrate.

Acknowledgements. Our thanks for the USTHB Fluid Mechanics Laboratory.
For allowing us to use Ansys Fluent 14.

References

1. Pawlowski L (2008) The science and engineering of thermal spray coatings, 2nd edn. John Wiley & Sons Ltd, Chichester, UK
2. Fauchais P, Montavon G, Vardelle M, Cedelle J (2006) Developments in direct current plasma spraying. Surf Coat Tech 201:1908–1921
3. Chandra S, Fauchais P (2009) Formation of solid splats during thermal spray deposition. J Therm Spray Technol 18:148
4. Li CJ, Li CX, Yang GJ, Wang YY (2006) Examination of substrate surface melting induced splashing during splat formation in plasma spraying. J Therm Spray Technol 15(4):717–724
5. Goutier S, Vardelle M, Fauchais P (2012) Understanding of spray coating adhesion through the formation of a single lamella. J Therm Spray Technol 21:522
6. Kang CW, Tan JK, Pan L, Low CY, Jaffar A (2011) Numerical and experimental investigations of splat geometric characteristics during oblique impact of plasma sparying. Appl Surf Sci 257:10363
7. Tabbara H, Gu S (2012) Modelling of impingement phenomena for molten metallic droplets with low to high velocities. Int J Heat Mass Transf 55:2081
8. Oukach S, Hamdi H, El Ganaoui M, Pateyron B (2015) Numerical study of the spreading and solidification of a molten particle impacting onto a rigid substrate under plasma spraying conditions. Therm Sci 19:277–284
9. Zhang H (1999) Theoretical analysis of spreading and solidification of molten droplet during thermal spray deposition. Int J Heat Mass Transf 42(14):2499–2508
10. Bussmann M, Chandra S, Mostaghimi J (2000) Modeling the splash of a droplet impacting a solid surface. Phys Fluids 12(12):3121–3132

11. Pasandideh-Fard M, Chandra S, Mostaghimi J (2002) A three-dimensional model of droplet impact and solidification. Int J Heat Mass Transf 45:2229
12. Jiang XY, Wan YP, Wang XY, Zhang H, Goswami R, Herman H, Sampath S (2000) Investigation of splat/substrate contact during Molybdenum thermal spraying. In: Berndt CC (ed) Thermal spray: surface engineering via applied research, ASM International, Materials Park, pp 729–36
13. Wang SP, Wang GX, Matthys EF (1999) Deposition of a molten layer of high melting point material: substrate melting and solidification. Mater Sci Eng, A 262:25–32
14. Zarzalejo LJ, Schmaltz KS, Amon CH (1999) Molten droplet solidification and substrate remelting in microcasting—part I: numerical modeling and experimental verification. Heat Mass Transf 34:477–485
15. Zhang H, Wang XY, Zheng LL, Jiang XY (2001) Studies of splat morphology and rapid solidification during thermal spraying. Int J Heat Mass Transf 44(24):4579–4592
16. Li L, Wang XY, Wei G, Vaiday A, Zhang H, Sampath S (2004) Substrate melting during thermal spray splat quenching. Thin Solid Films 468:113–119
17. Rajesh KS, Arvind K (2015) Substrate melting and re-solidification during impact of high-melting point droplet material. J Therm Spray Technol
18. Brackbill JU, Kothe DB, Zemach C (1992) A continuum method for modeling surface tension. J Comput Phys 100:335
19. Voller VR, Prakash CA (1987) Fixed-grid numerical modeling methodology for convection-diffusion mushy region phase-change problems. Int J Heat Mass Transf 30:1709–1720
20. Aziz SD, Chandra S (2000) Impact, recoil and splashing of molten metal droplets. Int J Heat Mass Transf 43:2841–2857
21. Wang SP, Wang GX, Matthys EF (1998) Melting and resolidification of a substrate in contact with a molten metal: operational maps. Int J Heat Mass Transf 41(10):1177–1188
22. Keshri AK, Agarwal A (2011) Splat morphology of plasma sprayed aluminum oxide reinforced with carbon nanotubes: a comparison between experiments and simulation, Surf Coat Technol

Shearing of Aluminium Rods for the Production of Billets for Bulk Metal Forming Operations

Bernd-Arno Behrens(✉), Kai Brunotte, and Lennard Lippold

Institute of Forming Technology and Machines (IFUM), Leibniz Universität Hannover, An der Universität 2, 30823 Garbsen, Germany
{behrens, brunotte, lippold}@ifum.uni-hannover.de

Abstract. Shearing of bulk material is a cost efficient method of producing billets for subsequent forging operations. The procedure of rod shearing is characterised by high billet output quantities without metal removal. This technique is widely used in industrial processing of steel materials. Aluminium billets however, are mostly produced by sawing, causing higher manufacturing costs due to comparably low output. The aim of this paper is an enhancement of sheared aluminium billet quality in order to fulfil industrial standards required for further forging operations. To enhance the quality of sheared aluminium billets, the shearing process parameter clearance (distance between the blades) was varied to create billets of four different aluminium alloys, as EN AW 5754 (UNS A95754), EN AW 5083 (UNS A95083), EN AW 6082 (UNS A96082) and EN AW 7075 (UNS A97075) in different conditions. The quality of sheared billets is assessed by analysing inclination angle, waviness and surface defects of the sheared parts with 3D-measurement. In the investigation, it can be shown that the tensile strength of the investigated alloys influences the billet surfaces. High tensile strength results in positive billet quality. Further, small clearances show best results concerning inclination angles, while higher clearances result in decreased waviness.

Keywords: Shearing · Aluminium · Forging · Billet production · Shear stress

1 Introduction

Aluminium is a lightweight material with outstanding properties such as high specific strength, excellent formability, good machining properties and recyclability. This is the reason for the steady increase of demand for aluminium products of different production processes [1].

Forging represents a production technique for parts of superior mechanical properties. Forging processes are characterised by high productivity due to high grades of automation, which results in large quantities of forged parts [2]. Due to the large quantities, the production of billets for forging operations must fulfil high standards concerning productivity and quality.

F. Chaari et al. (eds.), *Advances in Materials, Mechanics and Manufacturing*,
Lecture Notes in Mechanical Engineering,
https://doi.org/10.1007/978-3-030-24247-3_15

Shearing represents an economic billet production method, since it creates high billet output with short process times and a minimum of scrap material. Billets of round, rectangular or tube shaped profiles can be processed in shearing procedures [3].

The schematic illustration of a commonly used tool set-up for shearing operations is shown in Fig. 1. The shearing machine creates an acceleration of a moving blade, which is moved along a stationary blade, causing a separation of the billet from the remaining rod [4].

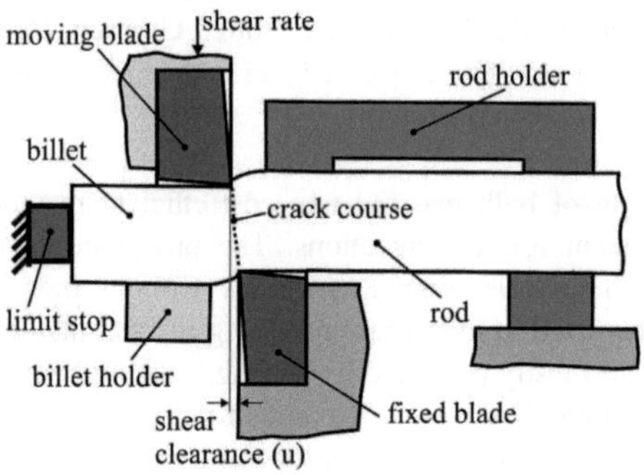

Fig. 1 Shearing operation [4]

The rod is clamped between a fixed and a moving blade. The moving blade is accelerated until contact to the rod is made, then causing elastic and plastic deformation of the material. The two blades cut into both sides of the rod. Further movement of the blade exceeds the deformability of the, not yet cut core rod material causing a fracture of the remaining rod. The resulting billet surface is characterises by a smooth cut and a rough fracture surface [4].

The shearing technique is well established for producing billets of steel materials while aluminium alloy billets are mostly produced by sawing due to insufficient quality of sheared billets. The sawing process produces billets of higher quality, is however less economic in comparison to shearing [5].

Sheared billets for further forging operations must fulfil certain quality standards. Table 1 shows typical characteristics used to describe billet quality.

Table 1 Defects of billets for forging operations [4]

Defect	Description
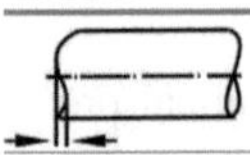	Waviness
	Inclination angle Ψ
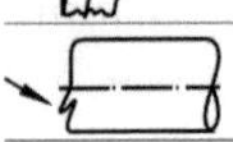	Surface defects

Since in most forging operations, billets are placed standing vertically between forging dies, the inclination angle of the billet should not exceed approx. 1° to avoid irregular material flow or even tool damage during forging [6]. Further, sheared surfaces with high waviness or surface defects can result in overlapping material in the finished forging parts. These overlaps can function as initiation for cracks or ruptures of the forged part in loaded conditions and should be avoided due to safety requirements [3].

Billet quality can be influenced through different process parameters of the shearing operation. The most relevant process parameter of shearing operations is the shear clearance, which describes the parallel distance between the blades [4]. Other approaches of increasing billet quality are high speed cutting with adiabatic effects during shearing [7] or shearing using compressive stress states [8].

2 Objective

In the current state of scientific and industrial knowledge, aluminium alloy billets of adequate quality required for following forging processes cannot be produced using shearing operations with process parameters for steel materials. Since shearing shows superior economical properties in comparison to sawing, the shearing process of aluminium is investigated with the aim of enhancing billet quality to a sufficient level. The most influential process parameter, shear clearance between the blades, is varied in a comprehensive experimental investigation of four industrially relevant aluminium alloys. To evaluate billet quality, the inclination angle, the waviness and the surface defects are determined.

3 Experimental Methods

Four aluminium alloys, shown in Table 2 are investigated in this shearing investigation. The used materials are initially characterised in tensile tests according to DIN EN ISO 6892-1:2017-02, while five specimens were used for the determination of mechanical properties as tensile strength and elongation at fracture. The selection of alloys and temper condition was chosen to display a wide range of different material properties. Preliminary work has shown that this selection of alloys in combination with the specific conditions are promising to create sheared billets of adequate quality.

Table 2 Aluminium alloys and temper conditions used in shearing investigations

Aluminium alloy	Condition
EN AW 5754—UNS A95754	H111
EN AW 5083—UNS A95083	H112
EN AW 6082—UNS A96082	T6
EN AW 7075—UNS A97075	O

Specimen (Ø = 20 mm) of the alloys are sheared with a shearing tool (Fig. 2) implemented in a screw press (Weingarten PSR 160) creating a cutting velocity of 430 mm/s in the shearing operation. By adjusting the distance between the blades in the setup, clearances of u = 0; 0.1; 0.2; 0.3 and 0.4 mm were implemented. For each experimental variation five specimens were used.

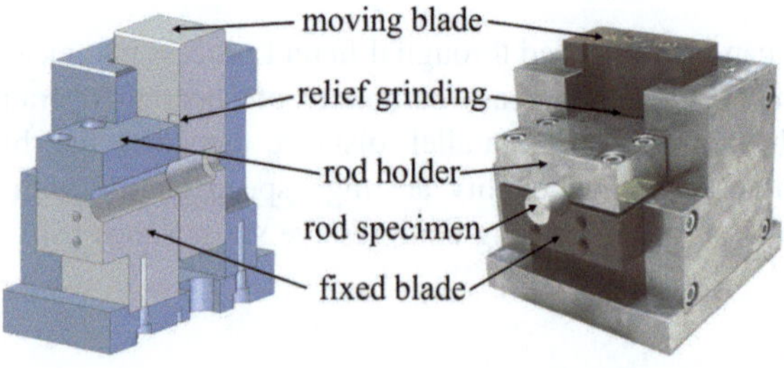

Fig. 2 Shearing tool used in the shearing operation

The sheared specimen surfaces were analysed with the 3D—measurement system Keyence VR-3200 as shown in Fig. 3. The surface quality was evaluated regarding surface waviness (w) and occurring surface defects such as ruptures or overlapping material. The inclination angle was measured with a digital inclination sensor with accuracy of ±0.1°.

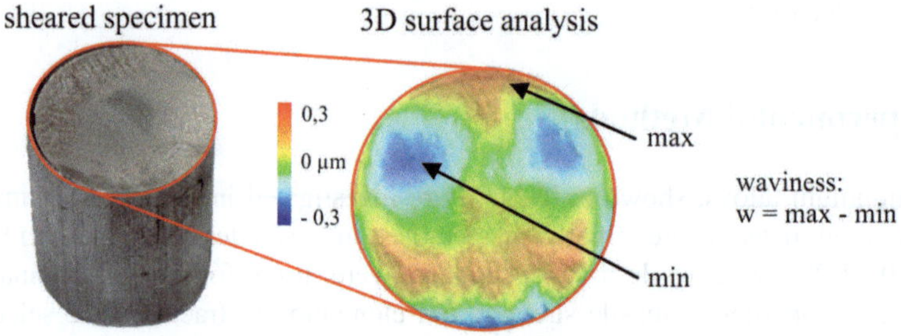

Fig. 3 surface analysis of sheared specimen

4 Results and Discussion

The determined stress-strain curves of the alloys are shown in Fig. 4, indicating significant differences in mechanical properties.

Comparably low ultimate tensile strengths of R_m = 236 MPa for EN AW 7075-O and R_m = 243 MPa for EN AW 5754-H111 were detected. While yield strengths are on an equal level, the elongation at fracture of EN AW 5754-H111 is distinctively higher, resulting in the highest ductility of all investigated materials.

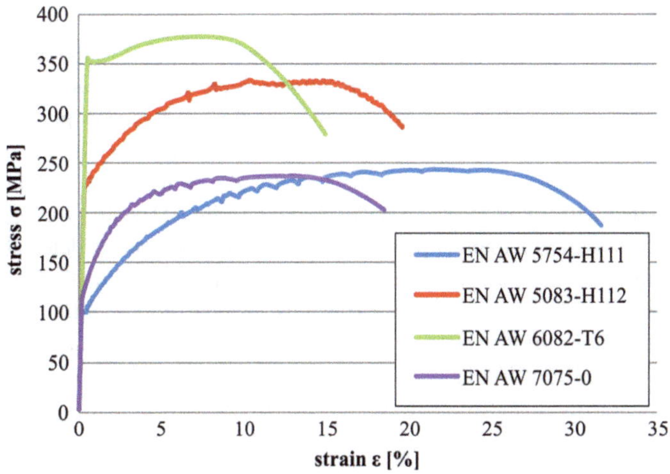

Fig. 4 Tensile strength of aluminium alloys

Higher ultimate tensile strengths of R_m = 332 MPa for EN AW 5083-H112 and R_m = 376 MPa for EN AW 6082-T6 were detected as well. In comparison to EN AW 5083-H112, EN AW 6082-T6 shows the highest yield strength and the lowest elongation at fracture of all investigated alloys, resulting in comparably low ductility.

In Fig. 5, surface height measurements of sheared billets of EN AW 5083-H112 are exemplified for each investigated clearance. The clearance (u) between the moving and the stationary blade was varied from 0 mm to 0.4 mm.

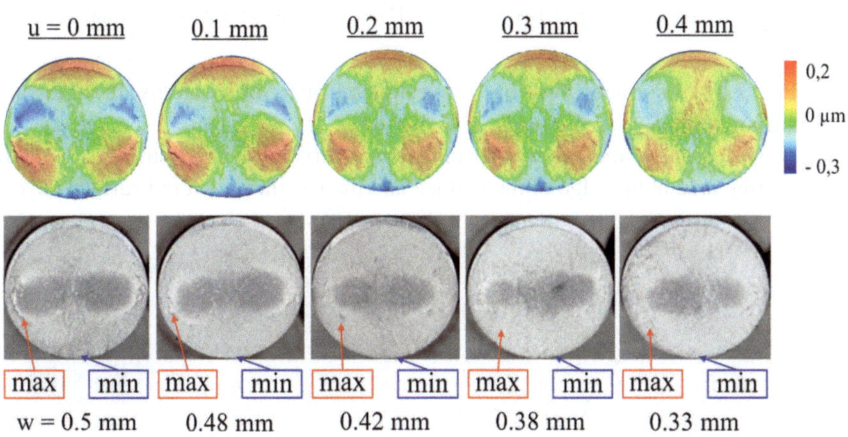

Fig. 5 Surfaces of sheared billets of EN AW 5083-H112 with varied clearance

While the cutting area is unaffected by the variation of process parameters, the topography of the fractured surfaces can be influenced through the variation of clearance. Surfaces with small waviness and low heights differences can be produced with the highest used clearance of 0.4 mm. This surface shows the smallest waviness and no identifiable surface defects. When decreasing the clearance, the waviness of the sheared surface increases. Further, surface defects such as elevation can be detected, which can lead to unintended material flow during further forging operations.

The measured inclination angle for each shear clearance is shown in Fig. 6 In comparison to the results of waviness, the best results in terms of inclination angles could be obtained with small clearances. As shown in Fig. 5, sheared surfaces of higher waviness can function as a contact point during vertical positioning, detectable in small inclination angles of the billets. Surfaces of low waviness lead to higher inclination angles. The highest elevation of these surfaces is located on the cutting surface while the fractured surfaces shows no significant elevations, leading to a high inclination angle during vertical positioning.

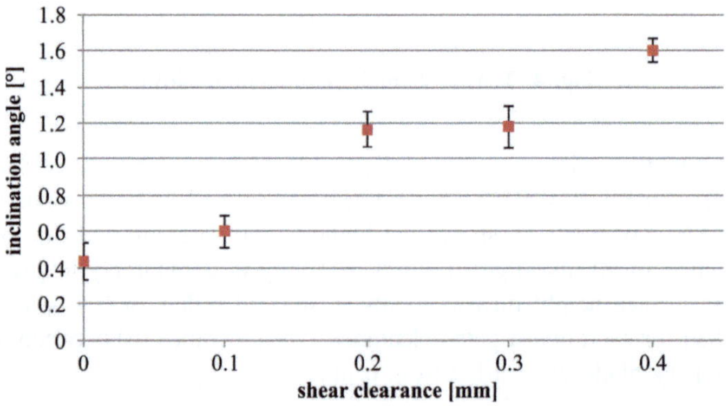

Fig. 6 Inclination angle of EN AW 5083-H112 as a function of shear clearance

On Fig. 7 billet surfaces from each investigated material sheared with a clearance of 0.4 mm are shown.

The surfaces show the process-specific characteristic sheared surfaces, with the smooth cutting area at the edge of the surface and the rougher fractured surface beyond.

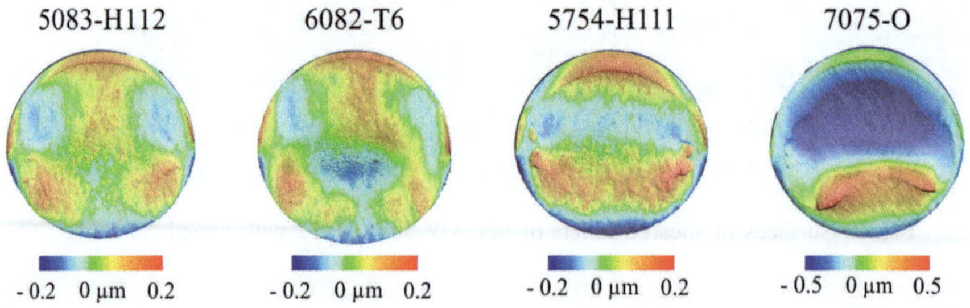

Fig. 7 Surfaces of sheared billets from different aluminium alloys with clearance of 0.4 mm

Sheared billets of alloys EN AW 5083-H112 and EN AW 6082-T6 showed small height differences of the fractured area with no identifiable surfaced defects. These surfaces can be rated as applicable for further forging operations, leading to forged parts with accurate surface properties. Small irregularities were detected on the surface of the billet of the material EN AW 5754-H111. These elevations must be assessed as negative since it may cause overlapping material in further forging operations. These may result in fractures of the forged parts and should be avoided to fulfil safety standards. Further, the specimen exhibit a significantly larger cutting area. This surface is created through the penetration by the sharp cutting blade, creating a smooth surface. Through further movement of the blade, the deformability of the material is exceeded and the fracture of the specimen occurs. Alloy EN AW 5754 shows the highest ductility of all investigated alloys and thus the highest deformability before rupturing, resulting in a larger smoothly cut surface area. Further the high ductility of the alloy leads to deformation of the profile resulting in ovality of the billet. Specimen of alloy EN AW 7075-O show the largest height differences leading to high surface waviness. The resulting billet surface quality is insufficient for any further forming operations. It was shown that adequate surface quality can be created by using the highest shear clearances 0.4 mm for aluminium alloys of high tensile strength. Alloys with lower strength show higher degrees of plastic deformations and result in surfaces with large defects and height differences.

The waviness was determined for each investigated aluminium alloy and each shear clearance. The results of the waviness measurements are shown on Fig. 8. For improved visualisation, trend lines were added for each alloy.

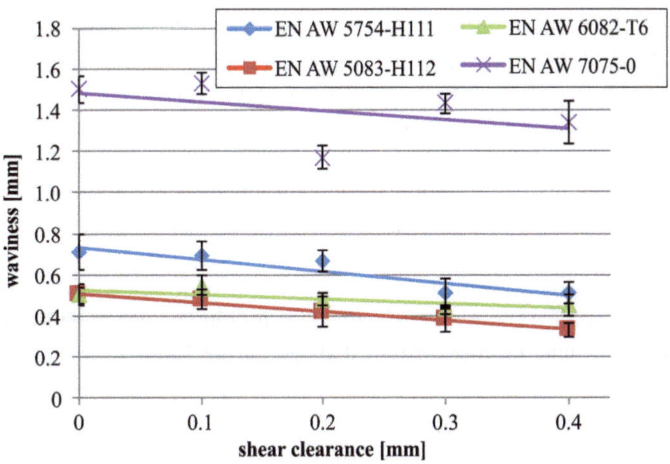

Fig. 8 Waviness of sheared billet surfaces for different shear clearances

The resulting waviness of all investigated alloys show a dependency on the used clearance, while the smallest values can be detected for the highest clearance of 0.4 mm for each material. The clearance between the blades strongly influences the prevailing stress state of the rod specimen leading to alterations of the fracture behaviour. The smallest waviness was found for alloys with high tensile strengths EN AW 5083-H112 and EN AW 6082-T6. The waviness of alloy EN AW 7075-O is on the highest level of all investigated alloys. The high values can be explained with the height differences of the specimen surfaces, as described in Fig. 7.

Figure 9 shows the inclination angle of sheared billets with the smallest clearance of 0 mm in comparison to the respective tensile strength.

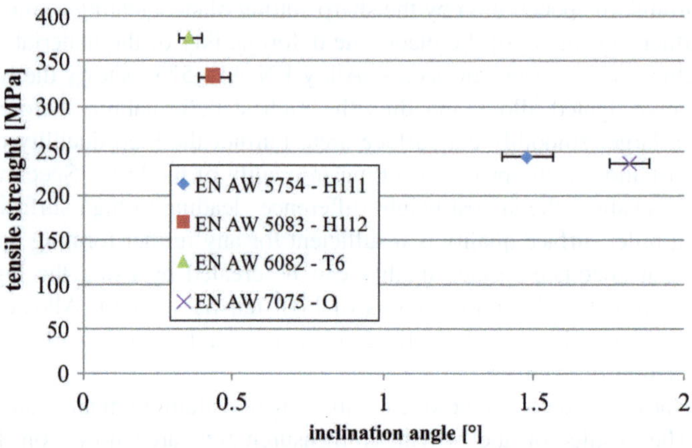

Fig. 9 Inclination angle depending on tensile strength of aluminium alloys (clearance 0 mm)

A correlation between tensile strength and inclination angle was determined. Shearing of aluminium alloys with comparably high tensile strengths, such as EN AW 6082-T6 (R_m = 376 MPa) and EN AW 5083-H112 (R_m = 332 MPa) results in excellent inclination angles of 0.36 and 0.44°. Specimen of alloys with lower tensile strength such as EN AW 5754-T111 (R_m = 243 MPa) or EN AW 7075-O (R_m = 236 MPa) show significantly higher inclination angles of 1.48 and 1.82°. Since the inclination angle of approx. 1° should not be exceeded for forging operations, billets from alloys such as EN AW 6082-T6 and EN AW 5083-H112 can be produced in shearing procedures with adequate waviness. Alloys of lower tensile strengths show higher inclination angles leading to inaccurate positioning between the forging dies and thus to insufficient material flow or tool damage during forging.

5 Conclusion

The mechanical properties such as tensile strength strongly influence the resulting billet quality. In the case of alloys with comparably high tensile strength, such as EN AW 6082-T6 and EN AW 5083-H112 billets of high quality concerning waviness, inclination angle and surface defects could be produced by varying shearing clearance. While aluminium billets with small surface waviness can be produced by shearing with comparably high shear clearances of 0.4 mm, small clearances of 0 mm create billets with smaller inclination angles. The ideal process parameters for the production of billets depend on the quality requirements for further forging operations. In the case of alloys with comparably high tensile strengths, an optimum of waviness and inclination angle could be found for a shear clearance of 0.3 mm, creating billets with comparably low waviness and low inclination angles as required for forging operations. Shearing specimen of alloys with low values of tensile strength as EN AW 5754-H111 or EN AW 7075-O lead to insufficient quality of inclination angle and surface waviness.

Acknowledgements. The presented investigations are results of the project "Investigations of the shear behaviour of aluminium alloys" (15868206), sponsored by the German Research Foundation (DFG). We are grateful for the assistance provided.

References

1. Polak S, Kaczyński P, Gronostajski Z, Jaskiewicz K, Krawczyk J, Skwarski M, Zwierzchowski M, Chorzępa W (2017) Warm forming of 7075 aluminum alloys. Procedia Eng 207:2399–2404
2. Behrens B-A, Lippold L, Kazhai M, Bouguecha A, Vucetic M, Hübsch C, Möhwald K (2016) Prediction and detection of wear mechanisms on an industry-oriented hot forging die. Adv Mater Res 1140:91–98
3. Behrens B-A, Lippold L, Knigge J (2013) Investigations of the shear behaviour of aluminium alloys. J Prod Eng Res Devel 7:319–328
4. Behrens B-A (2014) Billet shearing. In: CIRP, Laperrière L, Reinhart G (eds) CIRP encyclopedia of production engineering. Springer, Berlin, Heidelberg
5. Oudin J, Ravalard Y (1978) The experimental and theoretical analysis of plastic flow during the closed-die cropping of rectangular bars. Int J Mech Sci 21:63–70
6. Hu CL, Chen Q, Zhao Z, Li JW, Li ZM (2018) Study on the pre-shearing cropping process of steel bars. Int J Adv Manuf Technol 97:783–793
7. ICFG (1982) Cropping of steel bar—its mechanism and practice. Doc. No. 3/82, International Cold Forging Group, Portcullis Press
8. Simionato M, Ghiotti A, Bruschi S (2008) Billet cropping numerical modelling: an approach based on inverse analysis. Int J Mater Form Suppl 1:33–36

Evaluation of AW-6082 Aluminium Bar Shearing Simulation

Sonda Moakhar[1(✉)], Hamdi Hentati[1], Maher Barkallah[1], Jamel Louati[1], Christian Bonk[2], Bernd-Arno Behrens[2], and Mohamed Haddar[1]

[1] Research Laboratory of Mechanics, Modeling and Manufacturing (LA2MP), National School of Engineers of Sfax, Sfax, Tunisia
sonda.moakhar@stud.enis.tn, {hamdi.hentati, bark_maher}@yahoo.fr, {jamel.louati,mohamed.haddar}@enis.rnu.tn

[2] Institute of Forming Technology and Machines (IFUM), Leibniz Universität Hannover, Garbsen, Germany
fem@ifum.uni-hannover.de

Abstract. Bar shearing is an important operation that supplies semi-finished billets to many metalworking processes such as stamping, extrusion and precision forging. Temperature rise and stress state variation during shearing have a great influence on material behavior and rupture mechanics. Consequently, accurate simulation of shearing requires a precise material modeling. The studied material is the AW-6082 aluminium alloy. This paper concerns principally the improving of shearing simulation by means of adequate modeling of ductile failure. The major contribution of this study is to present a relatively uncomplicated method to calibrate a decoupled damage model. To this purpose, the Hooputra ductile damage (HDD) model is selected since it reflects the influence of different stress states and temperature variations on the mechanical failure of the material. The triaxiality is considered as indicator of the stress state. The identification of the parameters of the damage model is based on a hybrid experimental-numerical analysis of three characterization tests, namely tension tests on smooth bars, tension tests on notched bars and shear tests. The obtained calibrated damage model is employed to simulate shearing. The fracture is simulated using the "element deletion" technique. Computed shearing results are eventually evaluated by comparing simulated force-displacement curve to experimental one.

Keywords: Bar shearing · Decoupled damage model · Stress triaxiality · FEM

1 Introduction

Designing metalworking processes using finite element method (FEM) is nowadays a reliable method to optimize the process parameters and reduce testing costs. The exactness of process simulation is generally dependant from the accuracy of material modeling. In this context, an appropriate simulation of shearing requires the understanding of different phenomena occurring during the process and influencing material

F. Chaari et al. (eds.), *Advances in Materials, Mechanics and Manufacturing*, Lecture Notes in Mechanical Engineering,
https://doi.org/10.1007/978-3-030-24247-3_16

behavior and failure. Temperature increase and stress state alteration in the shear zone are the most important observed phenomena in aluminium bars shearing. Temperature can increase to 120 °C in aluminium shearing due to the plastic work in the shear zone [6]. Thus, temperature dependency is indispensable in damage modeling. Furthermore, various stress states are involved in shearing: compression, shearing and traction [9]. Stress triaxiality describes the local stress state and is one of the most important variables that must be considered in modeling aluminium ductile rupture [7]. That means that a wide range of triaxiality should be considered in aluminium failure [2]. Many damage models take account of the role of triaxiality in the calculation of damage variables [1, 12, 13, 16, 17]. Damage models are commonly classified into coupled and decoupled models. Coupled models allow for depicting the experimentally verified interaction between material behavior and damage evolution. However, the identification of their high number of parameters is too complicated and could reduce the accuracy of the obtained models [4]. Decoupled models, offer the advantage of being generally easy to implement and having few parameters to identify. However, they often prove unsuitable for complicated loading paths and large plastic deformations.

The geometry of sheared billets plays an important role in the results of some forging processes such as precision forging and stamping. It can induce at worst to forging defects such as underfill and folds [10].

The idea of this work is to present a relatively uncomplicated method to predict the rupture mechanics for the purpose of simulating accurately aluminium bars shearing. The selected damage model is the decoupled Hooputra ductile damage (HDD) model [11]. The calibration is based on three characterization tests with different triaxiality values, namely tension tests on smooth and notched bars and shear tests, which are performed in two temperatures. The identification of the parameters of the damage model is based on the combination of experimental and numerical analysis of these tests. The fracture strains are then calculated with the calibrated model and implemented in ABAQUS as a function of temperature and triaxiality to simulate shearing. A comparison between shearing simulation results and experimental results allows evaluating the calibration method.

2 Ductile Rupture Model

The HDD model was introduced by Hooputra et al. [11] and it assumes that the equivalent plastic strain at the onset of fracture is a two-term exponential function of triaxiality (Eq. 1).

$$\varepsilon_{eq}^{f} = ae^{-c\eta} + be^{c\eta} \tag{1}$$

where a, b, c are the material constants to identify and ε_{eq}^{f} is the equivalent plastic strain at the onset of fracture. η is the stress triaxiality, or triaxiality and is calculated as the ratio of hydrostatic stress (σ_m) and the von Mises equivalent stress (σ_{eq}) as follows:

$$\eta = \frac{\sigma_m}{\sigma_{eq}} \tag{2}$$

The damage is calculated as a cumulative variable (D), which is related to the evolution of strain (Eq. 3). In a simulation, the initiation of fracture occurs when the effective plastic strain exceeds the fracture effective plastic strain. This means that the damage variable D reaches a fixed value $D_{\max} = 1$.

$$D = \int_0^{\varepsilon_{eq}^f} \frac{d\varepsilon_{eq}}{\varepsilon_{eq}^f(\eta)} \tag{3}$$

where ε_{eq} is the equivalent plastic strain.

3 Numerical Parameters Identification

The identification of the HDD model constants for the simulation of shearing requires an appropriate selection of characterization tests involving a large range of triaxiality. Test temperature should also be varied in the observed range in the shear zone, as noted in Sect. 1. Three types of tests are carried out: round smooth bars traction (RST), notched round bars traction (NRT) and shear tests (ST). The material of the specimens is the aluminium alloy AW-6082. The tests were repeated for two temperatures: 20 and 100 °C. The strain rate is equal to 0.01 s^{-1} in all experiments. All experiments were performed at the Institute of Forming Technology and Machines (IFUM) at Leibniz Universität Hannover and are detailed in another publication of the authors [15].

The plasticity is tabular input in the material model according to flow curves generated from compressive tests. In this step, the ductile rupture is not modeled.

Equivalent plastic strains and triaxialities are numerically identified according to the following procedure. First, simulations of characterization tests are carried out using the finite element analysis software ABAQUS/Explicit. For smooth and notched round bars, axisymmetric specimens are modeled using the 4-node bilinear axisymmetric quadrilateral element (CAX4R). For the shear specimen, a 3D model is built with a thickness of 2 mm using the 8-node linear brick element (C3D8R). A fine mesh with an element size of 0.1 mm is applied in the rupture zone. Only the plastic behavior is implemented. Figure 1 shows the established geometries and mesh.

The numerical simulation of characterization tests ought to having similar results as experimental tests. To this end, simulation parameters, like mesh and increment sizes, are adjusted in order to obtain an acceptable similarity between simulated and experimental force-displacement curves in each test case. As an example, simulated and experimental force-displacement curves of smooth bars traction in 100 °C is presented on Fig. 2.

The average value of triaxiality over the deformation of every test is then calculated from the central element of the specimen. The equivalent strain at fracture is also determined for this element. The three experimental points are plotted on the fracture strain versus triaxiality plane for every test temperature (Fig. 3). To identify the

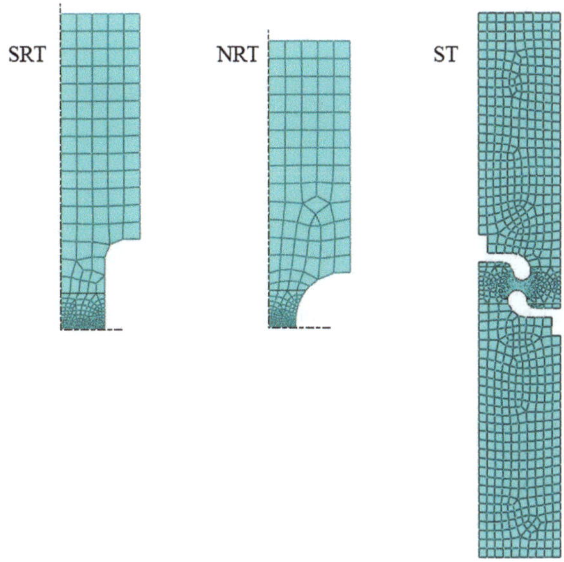

Fig. 1 Geometry and mesh of SRT, NRT and ST specimen models Adapted from Moakhar et al. [14]

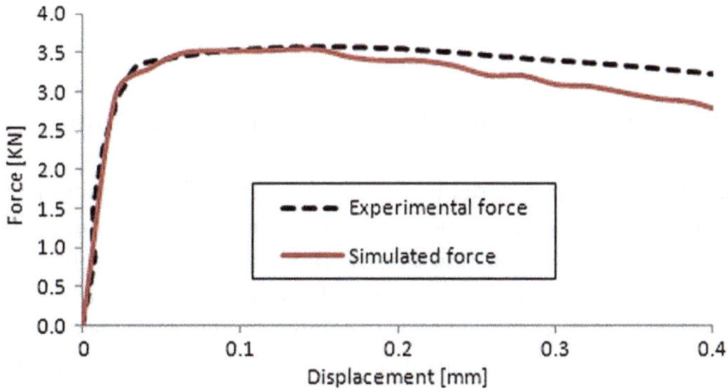

Fig. 2 Comparison of simulated force-displacement curve of the SRT test to the experimental one

parameters of the Eq. 1, a curve-fitting procedure is conducted on Matlab. The obtained correlation coefficients of the fracture limit curves in 20 °C and in 100 °C are respectively 0.42 and 0.99. The first coefficient is moderate because the scattering of experimental points could not be fitted under consideration of the ordinary shape of fracture limit curves. It should be noted that three experimental points for every curve remain small for fitting procedure. A larger number of experiments involving other triaxiality values would improve the exactness of the calibrated damage model. On the other hand, the experiments in 100 °C are well fitted with the HDD model. These

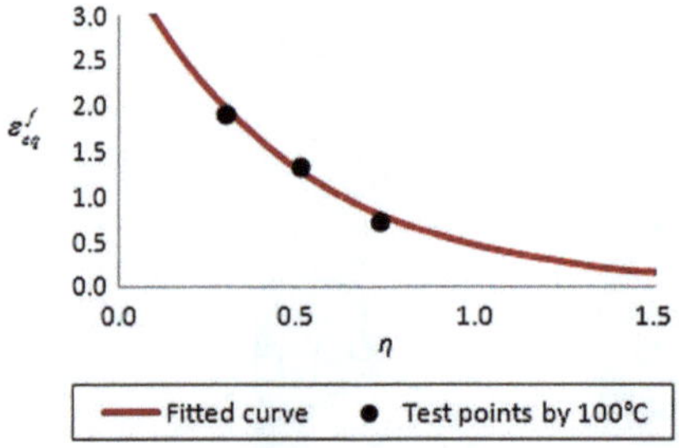

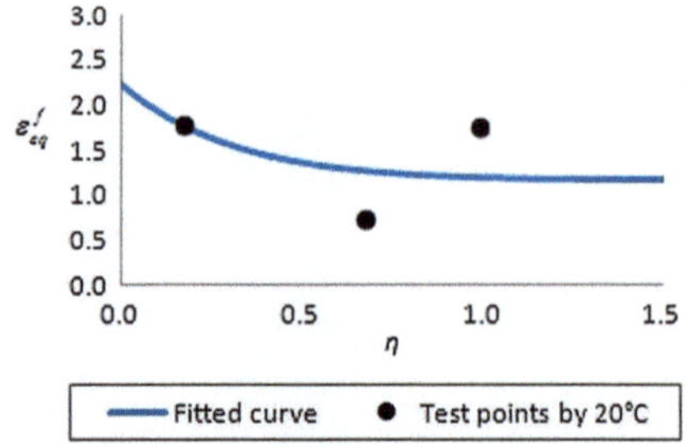

Fig. 3 Experimental points of characterization tests and fitted fracture limit curves in 100 °C (left) and 20 °C (right)

curves show that the increase in temperature to 100 °C reduces the strain at fracture by elevated triaxiality. Secondly, the strain at fracture in both temperatures is a decreasing function of triaxiality.

The obtained calibrated HDD equations in 20 and 100 °C are given respectively in Eqs. 4 and 5:

$$\varepsilon_{\text{eq}}^{\text{f}} = 1.12e^{-3.13\eta} + 0.03e^{3.13\eta} \tag{4}$$

$$\varepsilon_{\text{eq}}^{\text{f}} = 1.83e^{-2.12\eta} - 2e^{2.12\eta} \tag{5}$$

4 Evaluation of Aluminium Bar Shearing Simulation

A three dimensional shearing model is simulated in ABAQUS/Explicit using plasticity law as described in the previous section. Material behaviour at fracture, which consists in equivalent plastic strains as a function of triaxiality and temperature, is generated using Eqs. 4 and 5 and introduced in a tabular form under the ductile damage tab. The model components as well as the triaxiality evolution in the shear zone are presented on Fig. 4, where *d* is the tool displacement in mm. Different triaxiality values are met in the shear zone indicating different stress states from compression (negative triaxiality) to tension (positive triaxiality) with a dominance of shear stress state (nearly zero triaxiality). Hence, a damage modelling considering the influence of triaxiality on rupture mechanics would be appropriate.

Therefore, the negative triaxiality states should be covered by further tests that involve compression stress state. The results of these tests would be then integrated in the damage model calibration.

In order to evaluate the material modelling, simulation results and experimental ones are compared. Figure 5 shows the force-displacement curves from the FE simulation and from the experimental shearing test (experimental data were obtained from a previous study of [3]. It can be clearly seen here that the maximal shear force is correctly predicted. Therefore the simulated force decreases prior to the measured force. This can be due to the combined ductile and shear damage modes that occur

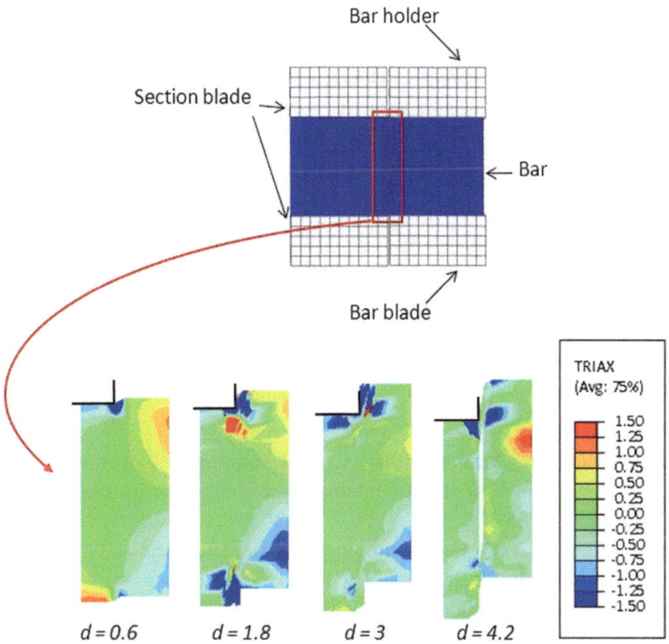

Fig. 4 Simulation model of shearing and triaxiality evolution in the shear zone

during shearing [8]. However, the utilized damage model considers only the ductile mode. The numerical results could also be improved by a suitable element deletion technique that better represents crack propagation [5]. As shown on Fig. 4., deleted elements induce volume loss which results in lower friction between the two bar sections. Consequently, simulated shearing force is expected to decrease sooner as expected. Mesh refinement could be a solution for such problem but it provokes considerably longer calculating time.

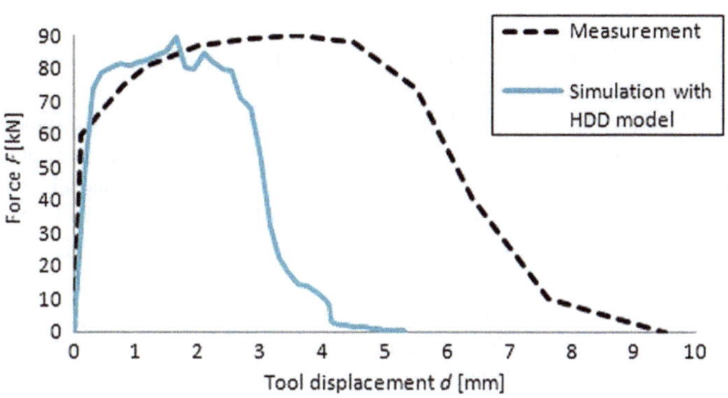

Fig. 5 Force-displacement curves of simulated and experimental shearing Adapted from Moakhar et al. [14]

5 Conclusion

In this paper, aluminium bar shearing is numerically analysed. The material modelling for shearing simulation is carried out by means of a combined experimental and numerical method. The decoupled HDD model is thereby calibrated based on characterization tests involving different stress states and temperatures. Possible improvements of the numerical and experimental methods are suggested. This work is a part of a research about the improvement of shearing simulation based on adequate material characterization. An enhancement of the quality of sheared billets would be the next point of interest of this study in addition to the use of other damage models.

References

1. Ayada T, Higashino K, Mori K (1984) Central bursting in extrusion of inhomogeneous materials. In: Proceedings of the 1st international conference on technology of plasticity (ICTP) vol 1, pp 553–558
2. Bao Y, Wierzbicki T (2004) On fracture locus in the equivalent strain and stress triaxiality space. Int J Mech Sci 46:81–98
3. Behrens BA, Lippold L, Knigge J (2013) Investigations of the shear behaviour of aluminium alloys, Prod Eng—Res Dev (German Academic Society for Production Engineering (WGP))
4. Bourgeon L (2009) Etude et modélisation des mécanismes d'endommagement en forge à froid, Dissertation, École Nationale Supérieure des Mines de Paris
5. Chow CL, Lu TJ (1992) A comparative study of continuum damage models for crack propagation under gross yielding. Int J of Fract 53(1):43–75
6. Dodd B, Bai Y (2014) Introduction to Adiabatic Shear Localization, revised edn, Imperial College Press, London
7. Driemeier L, Brünig M, Micheli G, Alves M (2010) Experiments on stress-triaxiality dependence of material behavior of aluminum alloys. Mech Mater 42:207–217
8. Duvari S, Isbir S, Ngaile G, Altan T (2003) Optimization of tool design in hot shearing of billets for forging, ERC/NSM-03-R-09, Engineering Research Center for Net Shape Manufacturing
9. Gutknecht F, Steinbach F, Clausmeyer T, Tekkaya AE. (2015) Advanced material model for shear cutting of metal sheets. In: Proceedings of the XIII COMPLAS, CIMNE, Barcelona1, pp 70–181
10. Hawryluk M, Jakubik J (2015) Analysis of forging defects for selected industrial die forging processes. Eng Fail Anal 59:396–409
11. Hooputra H, Gese H, Dell H, Werner HA (2004) Comprehensive failure model for crashworthiness simulation of aluminium extrusions. Int J Crashworthines 9(5):449–464
12. Johnson GR, Cook WH (1985) Fracture characteristics of three metals subjected to various strains, strain rates, temperatures and pressures. Eng Fract Mech 21(1):31–48
13. Lemaitre J (1985) A continuous damage mechanics model for ductile fracture. J Eng Mater Technol 107:83–89
14. Moakhar S, Hentati H, Barkallah M, Louati J, Haddar M (2018) Damage model calibration for the shearing simulation. In: 2nd International conference on acoustics and vibration (ICAV' 2018), Hammamet, Tunisia

15. Moakhar S, Hentati H, Barkallah M, Louati J, Bonk C, Behrens BA., Haddar M (in press) Modeling of the ductile damage–Application for bar shearing. Materialwiss Werkst. https://doi.org/10.1002/mawe.201800128
16. Oyane M, Sato T, Okimoto K, Shima S (1980) Criteria for ductile fracture and their applications. J Mech Work Technol 4(1):65–81
17. Rice JR, Tracey DM (1969) On the ductile enlargement of voids in triaxial stress fields. J Mech Phys Solids 17:201–217

An Analytical Approach for Modeling a Multibody System During Pre-design with Application to the Railway System

Ghazoi Hamza[1(✉)], Maher Barkallah[1], Moncef Hammadi[2], Jean-Yves Choley[2], Alain Riviere[2], Jamel Louati[1], and Mohamed Haddar[1]

[1] Mechanics Modeling and Production Research Laboratory (LA2MP), National School of Engineers of Sfax (ENIS), University of Sfax, B.P. 1173, 3038 Sfax, Tunisia
hamza.ghazoi@gmail.com, bark_maher@yahoo.fr, louati.ttg@gnet.tn, mohamed.haddar@enis.rnu.tn
[2] QUARTZ EA 7393, SUPMECA, 3 rue Fernand Hainaut, 93407 Saint-Ouen Cedex, France
{moncef.hammadi, jean-yves.choley, alain.riviere}@supmeca.fr

Abstract. A multibody system can be considered as an assembly of interconnected components: flexible bodies, rigid bodies, joint and unit control. This paper describes an analytical methodology for the predesign of a multibody system mixing rigid and flexible structure and regarding the vibrational aspect. Indeed, one of the principal aspects that limit performance in mechanical systems is the presence of vibrations. This contribution is based on the object-oriented modelling with Modelica language. Structural engineers frequently encountered problems arising from deflection of flexible structure such as plate and beam under moving load. In this paper, we study the interaction between a dynamic exciter moving with a constant velocity over a flexible beam resting on an elastic foundation by using the theory of dynamic response of Euler Bernoulli beam. The study of the moving load is of great importance in the transportation field. This methodology will be illustrated to the railway system. In fact, the dynamic response of a railway track subjected to a moving train load may be simplified as a beam traversed by a moving load. An analytical based approach is considered for the dynamic simulation of the flexible multibody system. The effect of some parameters on the dynamic response of the system has been studied.

Keywords: Multibody system · Flexible structure · Predesign · Object oriented modeling · Analytical model

1 Introduction

A mechatronic system is an assembly of different components from different areas and different technology. The target of the preliminary design is to specify the characteristics of these components before defining them completely. At this level, to take into

F. Chaari et al. (eds.), *Advances in Materials, Mechanics and Manufacturing*,
Lecture Notes in Mechanical Engineering,
https://doi.org/10.1007/978-3-030-24247-3_17

account the design constraints, it is necessary to have simple models allowing the characterization of the component [1–3].

In the literature, there are several methodologies used for the predesign of a multibody system, taking into account the multi-physic constraints [4–6]. For example, Ghazoi et al. [7] proposed a new pre-designing method applied to a mechatronic system and regarding the vibrational effect. The vibration interaction between dynamics exciters (Motors) and receivers (Electronic cards) is studied. These components are located over a flexible structure (plate). The model has been developed using the objected oriented modeling language Modelica/Dymola.

The influence of the boundary conditions of a flexible beam on the dynamic response of a system is studied in [8]. The beam supports dynamic perturbations (motors) and receivers (electronic cards). Two types of beam boundary conditions are considered simply supported- simply supported and clamped-clamped. This methodology helps system architect to quickly build up system model, the characterization and the analysis of a mechatronic system at an early stage of design.

In another work, authors [9] create compacts analytical models which are flexible beams in Modelica/Dymola. These elements may be inserted in a mechatronic system in order to study the vibrational response at an early stage of design. This methodology is applied to a typical example of a mechatronic system which is the wind turbine. Some simulation results are presented in order to show the capability of this method to have an insight about the system dynamic behavior.

The problem of a continuously distributed structure (beam, plate, etc.) carrying a moving load has broad applications in engineering and mechanics such as flexible manipulators and highway bridges with moving vehicles [10].

In dynamic mechanical system modeling, initial focus requires defining a system boundary. The analysis of beam resting on elastic foundations is of interest several fields such as aero-space structure and foundation engineering. The study of the behaviors of structures subjected to moving loads has been the concern of several researchers [11–13]. For example, Frýba [11] studied the problem of a beam resting on Winkler foundation and traversed by a moving load using the Fourier transformation method to solve the governing differential equations.

In railway engineering the track structure is generally modeled as an infinite, flexible beam resting on continuous springs loaded by a moving force. Today, more sophisticated models are required due to the increase of vehicles speeds in order to take into account dynamic effects for track design [14, 15].

Our contribution aims to propose a new methodology for the predesign of a multibody system taking into account the vibration constraint. More precisely, we focus on the dynamic response of a flexible beam resting on an elastic foundation and subjected to a moving load at a constant speed.

The paper is structured as follows. Section 2 describes the mathematical model used in this study; Sect. 3 gives the implementation of the model in Modelica/Dymola. The railway system is used as a support study. Numerical results are provided in Sect. 4. Section 5 concludes the paper.

2 Description of the Mathematical Model

The ordinary Bernoulli Euler theory of beams is considered in this paper, where the effect of rotary inertia is neglected. Beam on elastic foundation subjected to a moving load. Where, the foundation has been modelled by springs with constant foundation stiffness (Fig. 1). The beam is considered of an infinite length. The material is elastic, homogeneous and isotropic. The effect of damping is neglected. The force moves from left end to the right end with a constant speed v_0 [16].

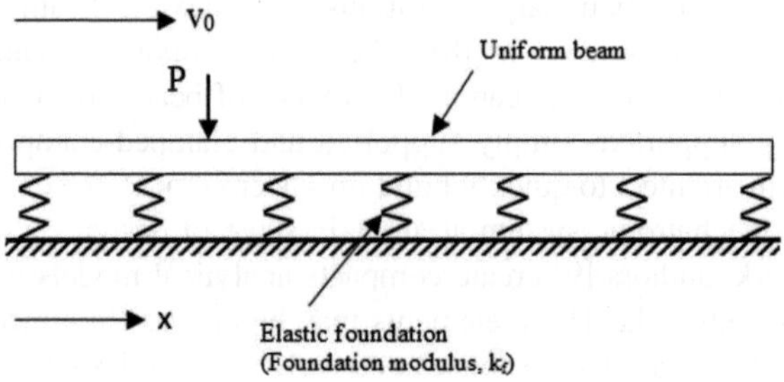

Fig. 1 Infinite beam on an elastic foundation subjected to an external moving load

The transverse vibration of the beam is governed by the following partial differential equation [16]:

$$EI\frac{\partial^4 w(x,t)}{\partial x^4} + \rho S\frac{\partial^2 w(x,t)}{\partial t^2} + k_f w(x,t) = f(x - v_0 t) \tag{1}$$

where w is the beam deflection at point x, E is the module of elasticity, I the moment of inertia of the cross- section, ρ is the density of material, S is the cross-sectional area, k_f is the foundation constant. $f(x - v_0 t)$ is the applied force moving with constant speed v_0

Defining the following,

$$z = x - v_0 t \tag{2}$$

Equation (1) can be written as:

$$EI\frac{\partial^4 w(z)}{\partial x^4} + \rho S v_0^2\frac{\partial^2 w(z)}{\partial t^2} + k_f w(z) = f(z) \tag{3}$$

The solution of Eq. (1) is [16]:

$$w(z) = -\frac{P}{i.4EI\sqrt{\beta}\sqrt{\alpha - \sqrt{\beta}}}e^{-i\sqrt{\alpha-\sqrt{\beta}}z} + \frac{P}{i.4EI\sqrt{\beta}\sqrt{\alpha + \sqrt{\beta}}}e^{-i\sqrt{\alpha+\sqrt{\beta}}z} \tag{4}$$

where,

$$\alpha = \frac{\rho S v_0^2}{2EI} \tag{5}$$

$$\beta = \frac{\rho^2 S^2 v_0^4}{4E^2 I^2} - \frac{k_f}{EI} \tag{6}$$

3 Application in Railway System

3.1 Modelica

Modelica is an object orientated equation based language for modeling of large and heterogonous physical systems. It has been used especially in the design of multi-domain engineering systems such as aerospace and automotive applications involving electrical mechanical, hydraulic and control subsystems.

3.2 Implementation into Modelica

This section deals with the preliminary sizing and the object oriented modeling of the railway system in the Modelica/Dymola environment. The graphic layout of the model developed with Dymola, a simulation environment based on the Modelica modeling language. The theory presented in Sect. 2 was implemented as components in Dymola. The schematic diagram of the system is given in Fig. 2.

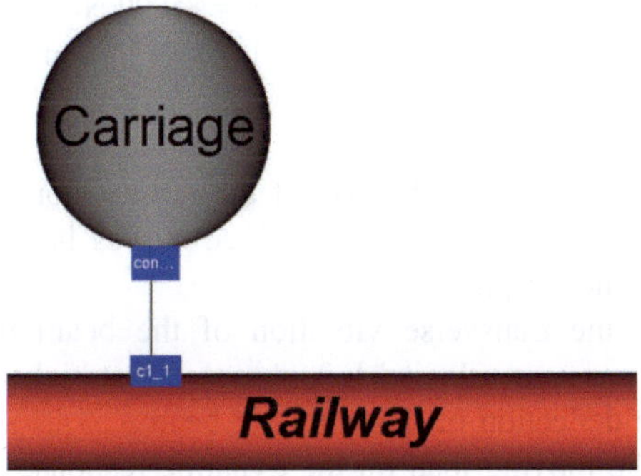

Fig. 2 Set up of the railway system model developed with Dymola

The system is composed of two components which are the railway model and the carriage model. Every component is represented by an icon.

The railway component includes the equation of motion and all other necessary equations to describe the beam. The railway model parameters include Young

Modulus, boundary conditions (the foundation stiffness) and the mass density, etc. The carriage model parameters include the load force and the speed.

Each component is equipped by a connector to be able to connect to other components. The code of the developed connector is the following:

```
connector C1
Modelica.SIunits.Position s;
Modelica.SIunits.Force F;
 a
end C1;
```

The developed connector is characterized by two variables, position and force.

4 Numerical Examples and Discussion of the Results

The effect of different parameters on the system response is exanimated. The material and geometric parameters of the system are shown in Table 1.

Table 1 System parameters

Parameters	Values
Mass density	$\rho = 7850\ \text{kg m}^{-3}$
Elastic modulus	$E = 2.10^{11}\ \text{N/m}^2$
Constant of velocity	$V_0 = 4000\ \text{m/s}$
The foundation stiffness	$K_f = 10^6\ \text{N/m}^3$
Force magnitude	$P = 85000\ \text{N}$
Moment of inertia	$I = 3.10^{-5}\ \text{kg.m}^2$

Figure 3 shows the transverse vibration of the railway for different points on the length of the beam. It can be seen that the three curves have the same oscillation frequency and they are not in phase.

Figures 4 displays the transverse vibration of the beam for various values of foundation stiffness at constant velocity. It is observed that higher values of foundation stiffness' k_f reduce the deflection of the vibrating beam.

The influence of the moving load on the dynamic response of the beam is investigated. Figure 5 shows the response of the beam for various values of the load. It can be seen that the dynamic deflection of the beam increases with the increase of the load force.

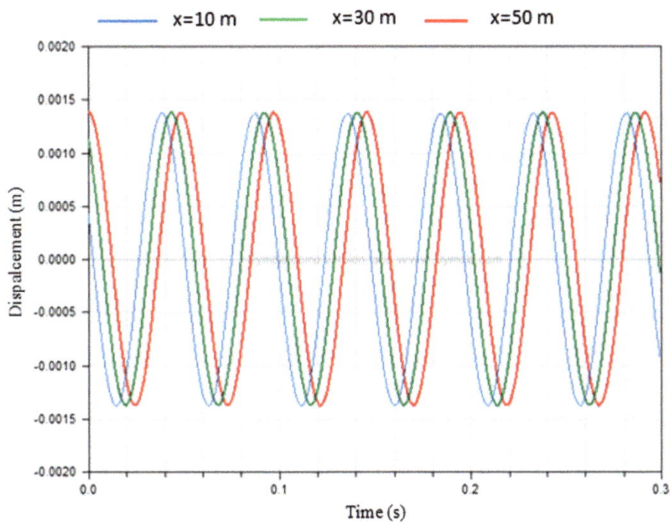

Fig. 3 Beam deflections for different points on the length of the beam

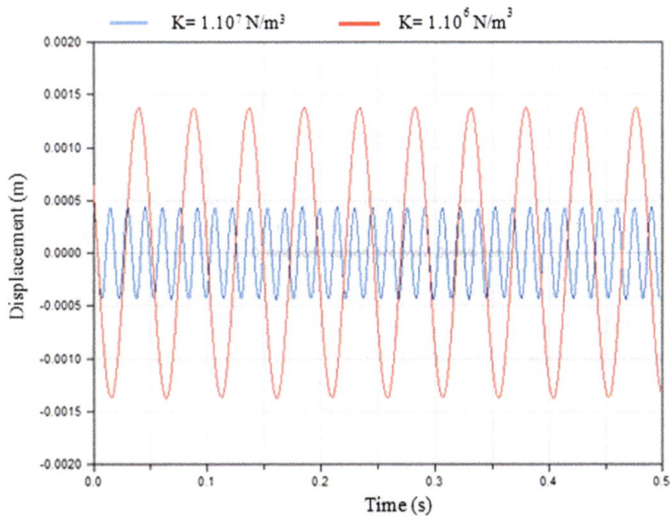

Fig. 4 Displacement response of the beam for various values of foundation modulus

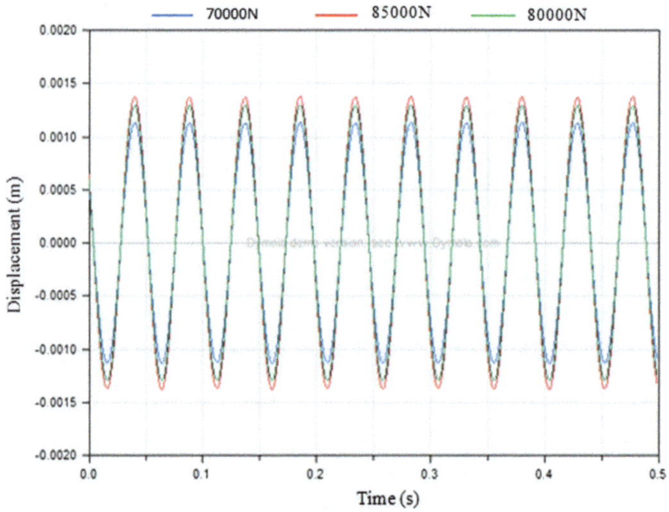

Fig. 5 Displacement response of the beam for various values of the load speed

5 Conclusion

This paper presents a new methodology for the preliminary design of a multibody system using the object oriented modeling methodology. This approach aims to study the vibrational behavior of a multibody system in the predesign phase. In fact, engineering machines during the life undergo multiple sources of vibrations. There are many engineering problems can be modeled as a coupled flexible beam- moving load. In this paper, we defined a Modelica model that can be used to predict the dynamic behavior of a beam under the action of a moving load. The beam is resting on an elastic foundation. The vibration response of a railroad track under the moving weight of the carriage can be determined using the present analysis. Several parameters having an impact on the vibrational behavior are analyzed. The modeling methodology is general enough that different boundary conditions of the beam can be implemented to create other models. The effect of damping is neglected.

References

1. Hamza G, Hammadi M, Barkallah M, Choley JY, Riviere A, Louati J, Haddar M (2018) Compact analytical models for vibration analysis in modelica/dymola: application to the wind turbine drive train system. J Chin Soc Mech Eng 39(1):121–130
2. Rzevski G (2003) On conceptual design of intelligent mechatronic systems. Mechatronics 13:1029–1044
3. Scheidl R, Winkler B (2010) Model relations between conceptual and detail design. Mechatronics 20:842–849

4. Hamza G, Choley JY, Hammadi M, Barkallah M, Louati J, Riviere A, Haddar M (2014) Analytical approach for the integrated preliminary analysis of mechatronic systems subjected to vibration, Mecatronics, In: 10th France–Japan/8th Europe–Asia Congress, Tokyo, Japan
5. Hamza G, Choley JY, Hammadi M, Barkallah M, Louati, J, Riviere A, Haddar M (2015a) Pre dimensioning of the dynamic properties of the wind turbine system using analytical approach, In: Design and modeling of mechanical systems II (CMSM '2015), Springer, pp 179–188
6. Hammadi M, Choley JY, Mhenni F (2014) A multi-agent methodology for multi-level modeling of mechatronic systems. Adv Eng Inform 28:208–217
7. Hamza G, Choley JY, Hammadi M, Riviere A, Barkallah M, Louati J, Haddar M (2015b) Pre-designing of a mechatronic system using an analytical approach with Dymola. J Theor Appl Mech 53:697–710
8. Hamza G, Hammadi M, Barkallah M, Choley JY, Riviere A, Louati J, Haddar M (2018) Conceptual design decision support of a mechatronic system using analytical approach with Modelica. Mech Ind 19:103
9. Hamza G, Hammadi M, Barkallah M, Choley JY, Riviere A, Louati J, Haddar M (2017) Conceptual design methodology for the preliminary study of a mechatronic system: application to wind turbine system. Mech Ind 18:413
10. Fischer M, Eberhard P (2014) Simulation of moving loads in elastic multibody systems with parametric model reduction techniques. Arch Mech Eng 61:209–226
11. Frýba L (2013) Vibration of solids and structures under moving loads, vol 1. Springer Science & Business Media
12. Lu S, Xuejun D (1998) Dynamic analysis to infinite beam under a moving line load with uniform velocity. Appl Math Mech 19:367–373
13. Mallik AK, Chandra S, Singh AB (2006) Steady-state response of an elastically supported infinite beam to a moving load. J Sound Vib 29:1148–1169
14. Wiberg J (2009) Railway bridge response to passing trains: measurements and FE model updating (Doctoral dissertation, KTH)
15. Jiang H, Bian X, Cheng C, Chen Y, Chen R (2016) Simulating train moving loads in physical model testing of railway infrastructure and its numerical calibration. Acta Geotech 11(2):231–242
16. Rao SS (2007) Vibration of continuous systems. John Wiley & Sons

Comparison Between the Effect of Magnetorheological Damper Force and a System Equipped with ADRC

Maroua Haddar[1(✉)], Riadh Chaari[1], S. Caglar Baslamisli[2], Fakher Chaari[1], and Mohamed Haddar[1]

[1] Mechanics, Modeling and Production Laboratory (LA2MP), Mechanic Department, National Engineering School of Sfax (ENIS), BP 1173, 3038 Sfax, Tunisia
{HADDAR.marwaGEM1, Fakher.chaari, Fakher.chaari}@gmail.com, Mohamed.haddar@enis.rnu.tn

[2] Department of Mechanical Engineering, Hacettepe University, 06800 Beytepe, Ankara, Turkey
Caglar.baslamisli@gmail.com

Abstract. The implementation of controller for a suspension system should have as low as possible complexity. The engineer should find the appropriate way to apply algorithms with low scaling parameters to achieve the required performances such as: ride comfort ride comfort, suspension spaces and dynamic tire load. For this reason, an appropriate way should be invented to apply the algorithm that gives the force generated by each actuator based on the motions of the vehicle which is received from various sensors located at different points of the vehicle. Two control strategies are investigated: First, the Active Disturbance Rejection Control (ADRC) control is investigated for showing its applicability to ameliorate ride comfort of passengers. Second, a traditional skyhook control scheme equipped with an Magnetorheological (MR) damper is investigated to show its superiority to give good performances of road holding and suspension space limits compared to the introduced intelligent controller ADRC. Furthermore, this semi-active controller is known as a simple control strategy with straightforward tuning process where only one gain parameter is needed for the implementation process. A simple prototype of quarter-car suspension is given to show the effectiveness of the introduced controllers. MATLAB/Simulink environment was used for investigating the comparison between the proposed techniques.

Keywords: Active control · Semi active control · ADRC · Ride comfort · Road

1 Introduction

The inconveniences of the passive control are harmful. For these reasons, some techniques such as semi-active control was developed. Currently, these improvements become intensified with a consideration of the arrival on the market of new intelligent

F. Chaari et al. (eds.), *Advances in Materials, Mechanics and Manufacturing*,
Lecture Notes in Mechanical Engineering,
https://doi.org/10.1007/978-3-030-24247-3_18

materials with modified properties. Consequently, according to the operational conditions, the characteristics of the mechanical items (damping, mass distribution, stiffness) are modified: for example switching the characteristics of dampers. Furthermore, there is no need of actuating energy. Therefore, there are no actuators in the sense of active force, only the shock absorber is required to modify the overall characteristics. Usually, a semi-active suspension can be actively remote by a control unit. The modification of damping coefficient can be continuously or discontinuously.

The control strategies for semi-active suspensions have been developed with various forms such as the Skyhook damping proposed by Karnopp [1]. It can be seen that semi-active suspensions operate under closed-loop control and are used as reference model [2]. Suspensions operate under closed-loop control and are used as reference model [2]. A variable damper effort is introduced in order to improve ride comfort in this type of controller [3] presented this kind of control in the case of quarter car model. The results give a better isolation to high frequency resonance compared to traditional passive system.

Active control is widely used in industrial applications where vibration levels are high. As opposed to passive control and semi-active control, active control requires external actuation or operative energy. This energy is related to the power of the actuators and amplifiers. Therefore, the control power and its effectiveness are limited only by the choice of the components and the control algorithms.

The present paper is focused on the concepts of 'Active Disturbance Rejection Controller" which are introduced by Gao et al. [4] and their application in the active suspension system. These tools have been presented in a large variety of academic applications and industrial [5]. The power produced by the actuator resides in estimating on-line unknown perturbations with a state observer. Only the knowledge of the input-output behavior of the plant is needed.

The organization of the paper is as follows. Section 2 describes the implementation process of ADRC controller. Section 3 gives a simple description of Skyhook control method. The comparison between numerical results is discussed in Sect. 4. Conclusion is summarized in the last section.

2 Intelligent Scheme of Control

The design of the ADRC suspension controller form proposed by Hasbullah et al. [6] is used in the present paper. The structure of ADRC controller is described as follows by two main parts: the extended state observer (ESO) used to estimate the internal and external disturbances, and the law of state feedback to generate the control signal of motion of active quarter car system presented in Fig. 1 are the following:

$$m_s\ddot{z}_s + d_s(\dot{z}_s - \dot{z}_u) + k_s(z_s - z_u) = u_A \tag{1}$$

$$m_u\ddot{z}_u - d_s(\dot{z}_s - \dot{z}_u) - k_s(z_s - z_u) + k_t(z_u - z_r) + d_t(\dot{z}_u - \dot{z}_r) = -u_A \tag{2}$$

m_s is the sprung masse, m_u is the unsprung masse, k_s is the suspension stiffness. the tire damping d_t is neglected because is very small. d_s is the suspension damping and k_t is

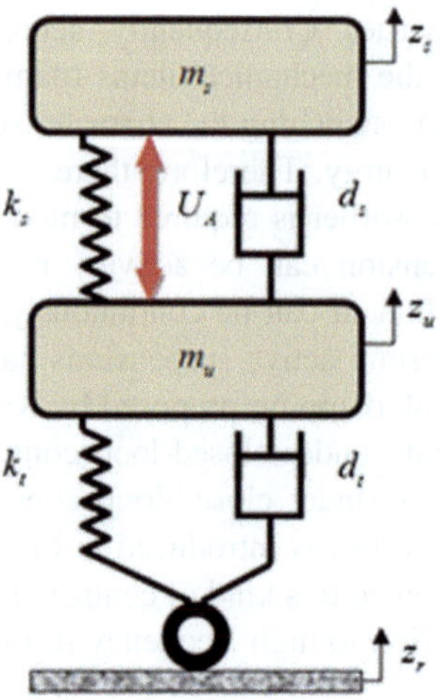

Fig. 1 Active quarter car system

the tire stiffness. The actuator force is denoted by u_A and to obtain this power, the following steps should be studied.

2.1 The System's State

The first step is the transformation of Eqs. (1) and (2) into state spaces system (3):

$$\begin{cases} \dot{y}_1 = y_2 \\ \dot{y}_2 = y_3 + \frac{1}{m_s} u_A \\ \dot{y}_3 = \dot{f}(t, y_1, y_2, w) \end{cases} \tag{3}$$

$y_1 = z_s$, $y_2 = \dot{z}_s$ and y_3 is all the un-known perturbations to be online estimated.

2.2 The Extended State Observer ESO

The key element of this active control scheme is the presence of an online observer to estimate all external and internal perturbations:

$$\begin{bmatrix} \dot{\hat{y}} \\ \dot{\hat{y}}_2 \\ \dot{\hat{y}}_3 \end{bmatrix} = \begin{bmatrix} L_1 & 1 & 0 \\ L_2 & 0 & 1 \\ L_3 & 0 & 0 \end{bmatrix} \begin{bmatrix} \hat{y}_1 \\ \hat{y}_2 \\ \hat{y}_3 \end{bmatrix} + \begin{bmatrix} 0 \\ \frac{1}{m_s} \\ 0 \end{bmatrix} u_A - \begin{bmatrix} L_1 \\ L_2 \\ L_3 \end{bmatrix} z_s \tag{4}$$

$[L_1, L_2, L_3]$ are the observer gains. All variables given by this symbol "^" represents the estimate of this state.

2.3 The Feedback Controller

The actuator force equation is the following:

$$u_A = -\frac{\hat{y}_3}{b_0} + \frac{1}{b_0}(K_p(y_{ref} - \hat{y}_1) + K_d(\dot{y}_{ref} - \hat{y}_2)) \tag{5}$$

K_p and K_d are respectively, the proportional gain and the derivative gain.

3 Skyhook Control

The skyhook configuration can be considered us the good scheme to achieve the ride comfort. For this reason, it is classified as is a comfort-oriented control strategy. In the research work of [7], an introduction of skyhook damper between the stationary sky and the sprung mass is chosen for the built of this configuration. The skyhook representation is given by Fig. 2.

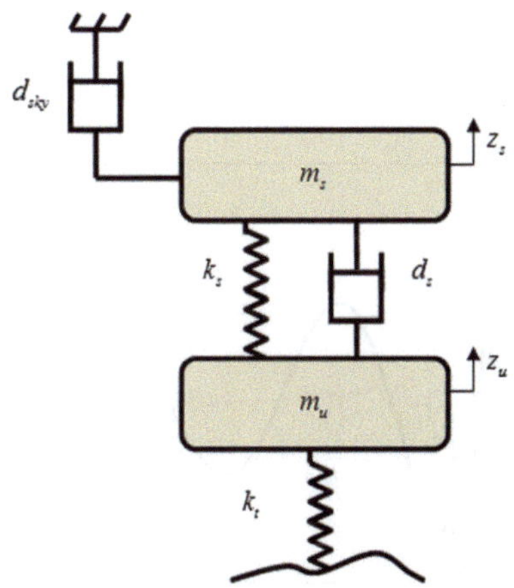

Fig. 2 Semi active quarter car system

The problem of skyhook damper is its inability to ameliorate handling performances.

For a semi-active suspension, the equations of motion is given by the following system:

$$m_s\ddot{z}_s + d_s(\dot{z}_s - \dot{z}_u) + k_s(z_s - z_u) - d_{sky}\dot{z}_s = 0 \tag{6}$$

$$m_u\ddot{z}_u - d_s(\dot{z}_s - \dot{z}_u) - k_s(z_s - z_u) + k_t(z_u - z_r) + d_t(\dot{z}_u - \dot{z}_r) = 0 \tag{7}$$

d_{sky} is the damping coefficient of the skyhook damper.

4 Numerical Results

The basic suspension parameters used in this paper are m_s = 285 kg, m_u = 41 kg, k_s = 17,756 N/m, k_t = 190,126 N/m, d_s = 535 N/(m/s) d_t = 0 N/(m/s). The road profile is chosen as a bump road given by the following equation:

$$z_r(t) = \begin{cases} 0.05(1 - \cos(8\pi t)), & t_1 = 0.5 \leq t \leq t_2 = 0.75 \\ 0 & otherwise \end{cases} \quad (8)$$

The three performance requirements of a quarter-car system can be summarized with the following aspects:

- **Ride comfort**: Fig. 3a shows sprung mass acceleration of the quarter car model. The ADRC controller has the best ride comfort against semi-active and passive controller. The vertical sprung mass acceleration Root Mean Square (RMS) improvement of active system reaches 80 and 75% for semi-active system (These values are calculated and compared with the passive responses). In the frequency domain (Fig. 3b), it is evident that the vibration of the sprung mass acceleration to disturbance is reduced immensely in active suspension strategy and is better in the

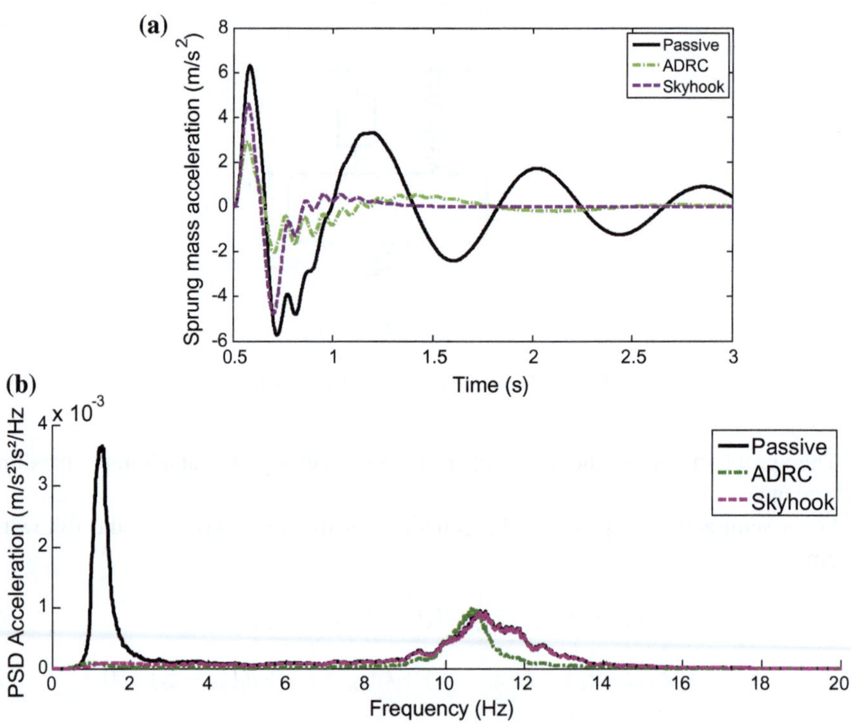

Fig. 3 **a** Time domain response of sprung mass acceleration and **b** frequency domain response of sprung mass acceleration

case of ADRC. It is clear that the first resonance peak is more reduced in the case of ADRC. However, the ride comfort given by semi active system is acceptable and can be adequate for operator requirement.

- **Suspension space limits**: The Skyhook control has the best ability of preventing the suspension from hitting its displacement limits (see Fig. 4a). It is able to give the best response with the lowest settling time. The high settling time in the case of ADRC can be related to the choice of PD gains. For getting the best ride comfort, the ADRC loses its ability to prevent the quarter car system from hitting its travel limits. This results are justified by the presence of high resonance peaks in the PSD curves of ADRC plotted in Fig. 4b. The controlled suspension spaces with ADRC presents a tradeoff between suspension requirements. However, the semi active suspension PSD response all fell into the acceptable ranges.
- **Road holding**: The good contact between the tyre and the road is given by Skyhook system (Fig. 5a). In fact, for improving the road holding quality by ADRC, we should make the proper choice of bandwidth parameter of observer ESO for getting a good disturbance rejection. From frequency curves depicted in Fig. 5b, the first peak of the accelerations of the sprung mass and the second peak of the unsprung

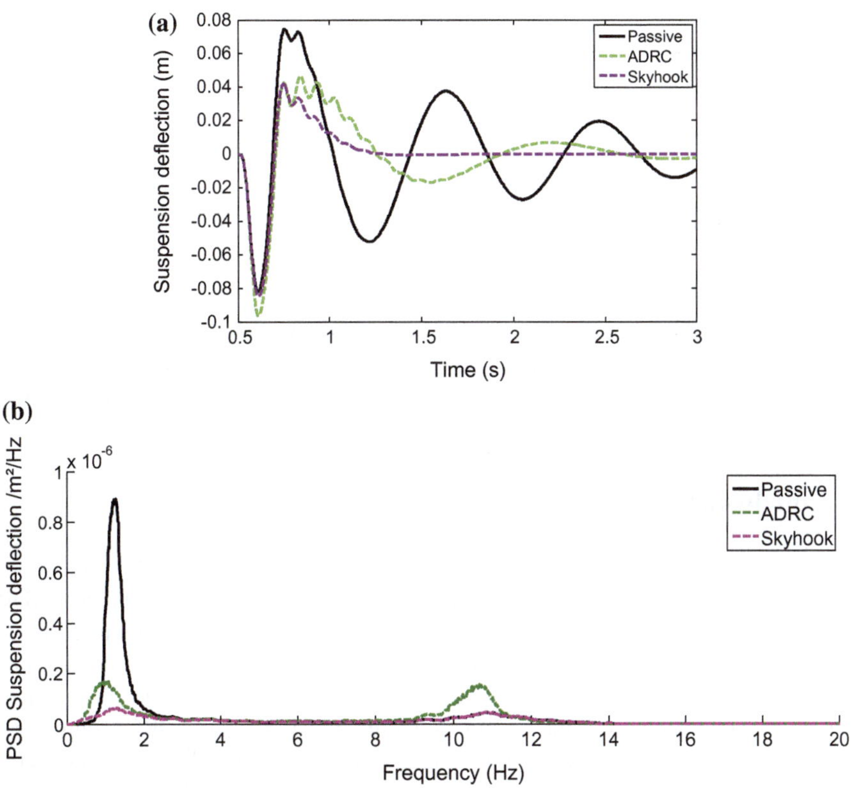

Fig. 4 **a** Time domain response of suspension deflection and **b** frequency domain response of suspension deflection

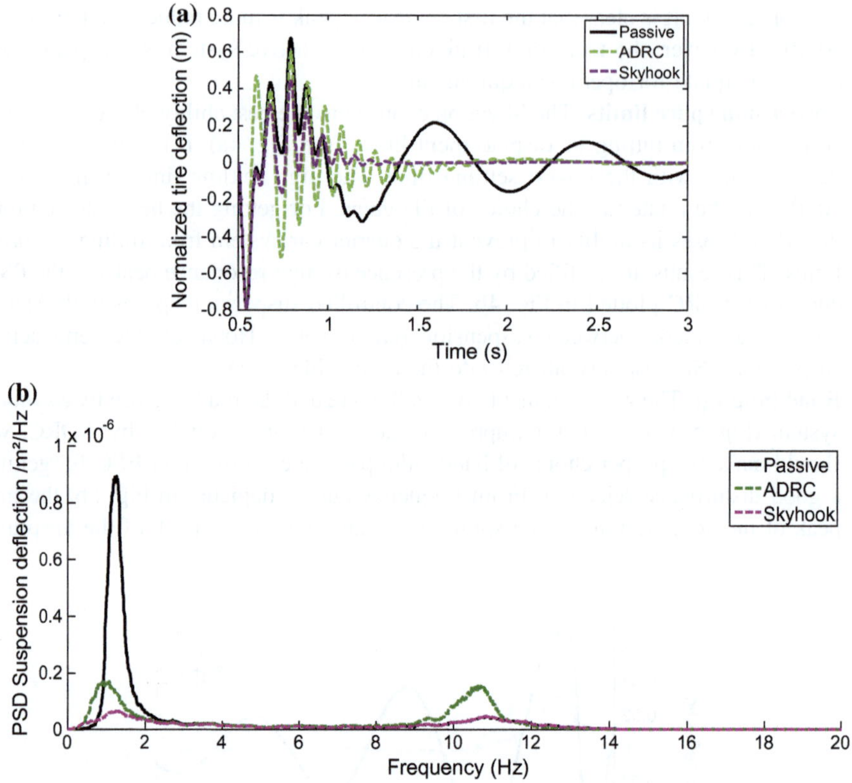

Fig. 5 **a** Time domain response of normalized tire deflection and **b** frequency domain response of tire deflection

mass (at their natural frequencies) decrease more in the case of Skyhook compared to ADRC. The high PSD level in ADRC curve can be justified by the presence of high disturbance perturbation transferred from the road to the suspension system.

5 Conclusion

This paper investigates a comparison between an intelligent controller and a semi active controller for the suspension system. From the depicted results, the ADRC shown its ability to ameliorate the ride comfort. However, the other required performances are not achieved. Therefore, this problem can be related to the optimal control of system that is not easy to be accomplished. On the other hand, the skyhook scheme with easy implementation and design is more effective than ADRC controller to ameliorate the suspension deflection response and the normalized tire deflection.

Furthermore, the tuning process of ADRC controller is not straightforward. The performance characteristics and the robustness of both semi-active and active suspension systems will be evaluated in the future publication.

References

1. Karnopp D (1995) Active and semi-active vibration isolation. In: Dans: current advances in mechanical design and production VI. s.l., Elsevier, pp 409–423
2. Zhang X et al (2013) Semi-active suspension adaptive control strategy based on hybrid control. s.l., s.n., pp 625–632
3. Faris WF, Ihsan SI, Ahmadian M (2009) A comparative ride performance and dynamic analysis of passive and semi-active suspension systems based on different vehicle models. Int J Veh Noise Vib 5:116–140
4. Gao Z, Huang Y, Han J (2001) An alternative paradigm for control system design, In: Proceedings of the 40th IEEE conference on decision and control, 2001. vol 5, pp 4578–4585
5. Herbst G (2013) A simulative study on active disturbance rejection control (ADRC) as a control tool for practitioners. Electronics 2(3):246–279. https://doi.org/10.3390/electronics2030246
6. Hasbullah F, Faris WF, Darsivan FJ (2015) Ride comfort performance of a vehicle using active suspension system with active disturbance rejection control. Int J Vehic Noise Vib 11:78–101
7. Karnopp D, Crosby MJ, Harwood RA (1974) Vibration control using semi-active force generators, Trans ASME, J Eng Ind 96:619–626

Geometrically Non-linear Free Vibrations of Simply Supported Rectangular Plates Connected to Two Distributions of Rotational Springs at Two Opposite Edges

Ahmed Babahammou(✉) and Rhali Benamar

University Mohammed V in Rabat, B.P. 765 Agdal, Rabat, Morocco
{ahmedbabahammou, rhali.benamar}@gmail.com

Abstract. Although the dynamic behavior of rectangular plates has been the subject of much research for many decades, it remains of a crucial importance in various engineering fields and some edge conditions have not yet been treated, especially those involving edges connected to distributed rotational springs and non-linear vibrations. Also, in the practice of Modal Testing, theoretical models are needed for quantitatively estimating the flexibility of the real plate supports. A complementary work is presented here corresponding to plates connected to a distribution of rotational springs at two opposite edges vibrating in the geometrically non-linear regime occurring at large vibration amplitudes. To build the plate trial functions, defined as products of beam functions in the x and y directions, the mode shapes of simply supported beams connected to rotational springs are first calculated. Then, after exposing the general formulation of the non-linear problem, based on Hamilton's principle and spectral analysis, the plate case is examined. Using the single mode approach, the backbone curves are determined, giving the non-linear frequency-amplitude dependence for plates having different combinations of stiffness and aspect ratios. It is noticed, as may be expected, that the obtained hardening non-linearity effect becomes more accentuated with increasing the rotational spring stiffness.

Keywords: Rectangular plates · Nonlinear vibration · Elastically restrained support · Hamilton principle · Mode shapes · Backbone curves · Single mode approach

1 Introduction

In spite of the amount of works performed on plate vibrations for many decades, only few papers deal with simply supported rectangular plates connected to two distributions of rotational springs at two opposite edges. Furthermore, to the knowledge of the authors, the geometrically nonlinear vibration of such elastically restrained plates has not been investigated, in spite of its theoretical and practical importance. On one hand, such edge conditions may be really encountered in practical situations. On the other hand, it should not be forgotten that the classical boundary conditions, i.e. simply supported and clamped, are practically impossible to achieve perfectly in real structures

F. Chaari et al. (eds.), *Advances in Materials, Mechanics and Manufacturing*,
Lecture Notes in Mechanical Engineering,
https://doi.org/10.1007/978-3-030-24247-3_19

since the supports have always some flexibility. Consequently, it is of a crucial importance for designers to be able and quantitatively estimate how far do the plate real dynamic characteristics deviate from the theoretical ones, corresponding to rigid vertical supports and completely free rotations (simply supported case, denoted in what follows as SS), and to rigid vertical supports and completely prevented rotations (clamped case, denoted as C). From the point of view of modal analysis and testing, the availability of theoretical results corresponding to flexible supports with various stiffness values may be very useful in interpreting experimental data provided by modal testing and also for an accurate identification process.

The purpose of this paper is to present the formulation of the problem of non-linear vibrations of simply supported rectangular plates connected to two distributions of rotational springs at two opposite edges, in both the linear and non-linear cases. The Rayleigh-Ritz method is used in the linear case, with plate functions defined as products of beam functions, with appropriate end supports, in each direction. The extension of the Rayleigh-Ritz method to the nonlinear case, developed and applied to various non-linear problems by Benamar and his co-authors [1–4] is used here to investigate the large vibration amplitudes of the plates examined.

Consider the plate shown in Fig. 1. It is supposed to be simply supported at the four edges and to be in addition connected to distributed rotational springs at the edges y = 0 and y = b. As mentioned above, the Rayleigh-Ritz method is used to investigate the linear vibration case, with plate functions defined as products of appropriate beam functions in each direction. The next section is concerned with a brief presentation of how to determine the mode shapes of the beam shown in Fig. 2.

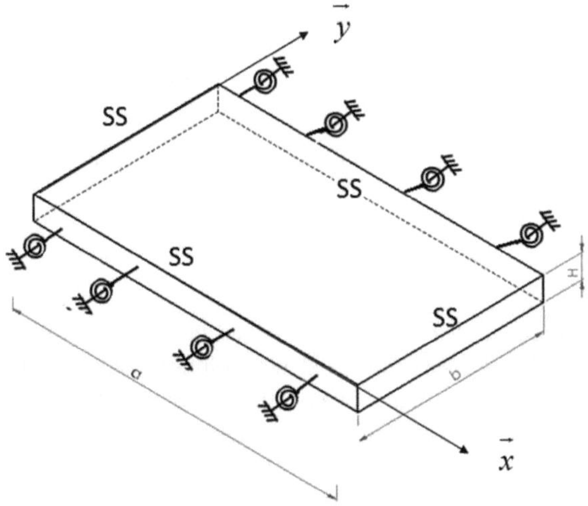

Fig. 1 SS Plate connected to distributed rotational springs at y = 0 and y = b

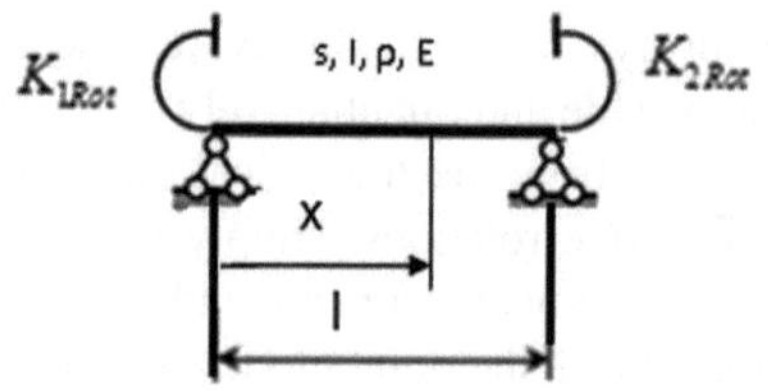

Fig. 2 SS beam, connected at the ends to rotational springs

2 Mode Shapes of a SS Beam Connected at the Ends to Rotational Springs

2.1 Theoretical Formulation

Consider the SS beam shown in Fig. 2, connected at the ends to two rotational springs of stiffness K_{1Rot} and K_{2Rot}. The beam has the characteristics indicated in the figure. The governing equation of the beam transverse vibration is governed by the well known differential equation [5]:

$$\frac{d^4w}{dx^4} + \beta^4.w = 0 \quad \text{with} \quad \beta^4 = \omega^2.\frac{\rho.S}{E.I} \tag{1}$$

The general solution of Eq. (1) can be written as:

$$w_i(x) = C_1\sin(\beta_i.x) + C_2\cos(\beta_i.x) + C_3\sinh(\beta_i.x) + C_4\cosh(\beta_i.x) \tag{2}$$

The β_i's are the beam mode shape parameters. C_1, C_2, C_3 and C_4 are determined by the end conditions:

$$\text{at}\, x = 0 : w = 0 \quad \text{and} \quad M_x = E.I.w'' = +K_{1Rot}.\frac{\partial w}{\partial x} \tag{3–4}$$

$$\text{at}\, x = l : w = 0 \quad \text{and} \quad M_x = E.I.w'' = -K_{2Rot}.\frac{\partial w}{\partial x} \tag{5–6}$$

M_x is the bending moment. Equations 3–6 give a linear system with 4 equations and 4 unknowns. To avoid having only the trivial zero solution, the determinant of the system must vanish, which gives the frequency equation, leading to the frequencies and mode shapes of the vibrating beam connected to the rotational springs. The Newton–Raphson algorithm was used to find the roots $\beta_1.l$ of the transcendental frequency equation, corresponding to the first mode and to various values of the rotational spring stiffness C*. The solutions are summarized in Table 1 and compared to previous results.

Table 1 Beam fundamental frequency parameters $\beta_1.l$ for different values of the rotational stiffness $C^* = \frac{K_{2Rot}.l}{E.I}$

C*	0	1	10	100	1000	
(a)	3.9237	4.0381	4.4229	4.6754	4.7151	4.7193
(b)	3.9237	4.0381	4.4229	4.6754	4.7151	4.7193

(a) Linear results obtained here (b) linear results obtained in Ref. [5]

3 Linear Vibration of SS Rectangular Plates Connected to Two Distributions of Rotational Springs at Two Opposite Edges

To examine the vibration of the plate shown in Fig. 1, the transverse displacement is assumed to be:

$$W(x, y, t) = w(x, y).\sin(\omega t) = a_i.w_i(x, y)\sin(\omega t) \tag{7}$$

In which the usual summation convention is used. The kinetic energy and the plate total strain energy V_T, which is the sum of the strain energy due to the bending V_b [6], plus the membrane strain energy due to the axial load induced by large deflections V_m [6], and the strain energy stored by the elastic edge restraints V_{spring} ($V_T = V_b + V_m + V_{spring}$) are given by Li [7]:

$$\begin{gathered} T = \frac{1}{2}\rho H \int_S \left(\frac{\partial W}{\partial t}\right)^2 .dS, \\ V_b = \frac{D}{2}\int_S \left[\left(\frac{\partial^2 W}{\partial^2 x} + \frac{\partial^2 W}{\partial^2 y}\right)^2 + 2.(1-\nu).\left(\left(\frac{\partial^2 W}{\partial x \partial y}\right)^2 - \frac{\partial^2 W}{\partial^2 x}.\frac{\partial^2 W}{\partial^2 y}\right)\right] dS \end{gathered} \tag{8–9}$$

$$\begin{gathered} V_m = \frac{3.D}{2.H^2}\int_S \left[\left(\frac{\partial W}{\partial x}\right)^2 + \left(\frac{\partial W}{\partial y}\right)^2\right]^2 .dS \\ V_{Spring} = \frac{1}{2}\int_0^a (K_{1Rot}.\left(\frac{\partial w(x,0)}{\partial y}\right)^2 + K_{2Rot}.\left(\frac{\partial w(x,b)}{\partial y}\right)^2 dx \end{gathered} \tag{10–11}$$

The basic spatial plate functions (Eq. 7), are defined as [2] $w_i(x, y) = P_I(x).Q_J(y)$, in which $P_I(x)$ and $Q_J(y)$ are beam functions with appropriate end conditions in each direction. The plate function index i is related to the indices I and J of the corresponding beam functions by: $i = N.(I-1) + J$, where N is the number of beam functions used. One obtains after discretization of the energy expressions [2]:

$$T = \frac{1}{2}\omega^2 a_i.a_j.M_{ij}.\cos^2(\omega t); \quad V_b = \frac{1}{2}a_i.a_j.K_{ij\_b}.\sin^2(\omega t); \tag{11–12}$$

$$V_m = \frac{1}{2}a_i.a_j.a_k.a_l.B_{ijkl}.\sin^4(\omega t), \quad V_{spring} = \frac{1}{2}a_i a_j K_{ij_{spring}}.\sin^2(\omega t). \tag{13–14}$$

In which M_{ij}, K_{ij_b}, B_{ijkl} and K_{ij_spring} are the mass, linear and non-linear rigidity tensors, defined by:

$$M_{ij} = \rho H \int_S w_i.w_j.dS \tag{15}$$

$$\begin{aligned} K_{ij\_b} = D \int_S \Bigg[&\left(\frac{\partial^2 w_i}{\partial x^2} + \frac{\partial^2 w_i}{\partial y^2}\right)\left(\frac{\partial^2 w_j}{\partial x^2} + \frac{\partial^2 w_j}{\partial y^2}\right) \\ &+ 2.(1-\nu).\left(\frac{\partial^2 w_i}{\partial x \partial y}.\frac{\partial^2 w_j}{\partial x \partial y} - \frac{\partial^2 w_i}{\partial x^2}.\frac{\partial^2 w_j}{\partial y^2}\right)\Bigg].dS \end{aligned} \tag{16}$$

$$B_{ijkl} = \frac{3D}{2.H^2}\int_S \left(\frac{\partial w_i}{\partial x}.\frac{\partial w_j}{\partial x} + \frac{\partial w_i}{\partial y}.\frac{\partial w_j}{\partial y}\right)\left(\frac{\partial w_k}{\partial x}.\frac{\partial w_l}{\partial x} + \frac{\partial w_k}{\partial y}.\frac{\partial w_l}{\partial y}\right).dS \tag{17}$$

$$K_{ij\_Spring} = \int_0^a \left(K_{1Rot}.\frac{\partial w_i(x,0)}{\partial y}.\frac{\partial w_j(x,0)}{\partial y} + K_{2Rot}.\frac{\partial w_i(x,b)}{\partial y}.\frac{\partial w_j(x,b)}{\partial y}\right).dx \tag{18}$$

The indices i and j are summed over 1, 2 … n, n being the number of the plate functions used ($n = N^2$).

A computer program has been written to calculate numerically the above parameters and solve the eigen value problem, corresponding to linear vibrations. To validate the program, a limit case, corresponding to C–SS–C–SS rectangular plates, for which the rotational spring stiffness tend to infinity, has been treated, for various plate aspect ratios α. The results obtained are summarized in Table 2 and compared to previously published results. Also, a comparison is made in Table 3 between the results obtained here and those given in [6], corresponding to the plate with elastic restraints shown in Fig. 1 ($\alpha = 0.4, 0.8, 1$). 36 plate functions have been used in the Rayleigh-Ritz formulation for various values of the spring stiffness. It appears that the percentage difference remains reasonably small for small and high values of the stiffness, corresponding to the SS and C edge conditions, and does not exceed 6.65% in all cases.

Table 2 Non-dimensional frequency parameters, of a C–SS–C–SS rectangular plate, for various plate aspect ratios α (a) results obtained here (b) results obtained in Ref. [4] (c) results obtained in Ref. [10]

	Plate non-dimensional frequency parameters									
α	0.2	0.33	0.4	0.5	0.6	0.66	0.7	0.8	0.9	1
(a)	10.358	11.372	12.149	13.698	15.74	17.2008	18.289	21.346	24.905	28.957
(b)	10.35	11.33	12.21	13.69	15.8	17.2	18.35	21.4	24.95	28.95
(c)	–	–	12.3	13.71	15.69	–	18.25	20.82	24.08	–

Table 3 Comparison of the frequency parameters $\frac{\omega^2.a^2.b^2.\rho}{\pi^4.D}$ (b) calculated here with (a) results given in [8] for different values of aspect ratio and different values of stiffness

$C^* = \frac{K_{iRot}.b}{D}$	First frequency parameters							
	α = 0.4		α = 0.6		α = 0.8		α = 1	
	(a)	(b)	(a)	(b)	(a)	(b)	(a)	(b)
0	8.41	8.41	5.138	5.1388	4.203	4.2035	4.000	4.000
0.5	8.438	8.4415	5.206	5.2081	4.326	4.3271	4.194	4.1941
1	8.468	8.4713	5.271	5.2736	4.441	4.4423	4.373	4.373
2	8.525	8.5261	5.387	5.3919	4.646	4.6483	4.691	4.691
3	8.572	8.5754	5.491	5.4958	4.825	4.8271	4.965	4.966
4	8.611	8.6199	5.584	5.5877	4.982	4.9837	5.205	5.2052
5	8.649	8.6604	5.666	5.6696	5.120	5.1219	5.415	5.4152
10	8.801	8.8169	5.974	5.9744	5.625	5.6252	6.169	6.1693
20	9.204	9.2167	6.671	6.672	6.719	6.723	7.777	7.7833
100	9.312	9.327	6.845	6.8452	6.981	6.9835	8.137	8.144
500	-	9.4382	6.974	7.0168	7.234	7.2382	8.494	8.505
∞	9.448	9.47	7.059	7.065	7.304	7.3091	8.593	8.605

4 Nonlinear Vibration of SS Rectangular Plates Connected to Two Distributions of Rotational Springs at Two Opposite Edges

To examine now the non-linear vibration of the rectangular plate examined, the membrane strain energy V_m induced by the large vibration amplitudes has to be taken into account in the application of Hamilton's principle as follows:

$$\delta \int_0^{\frac{2\omega}{\pi}} \left(T - \left(V_b + V_{spring} + V_m\right).dt = 0 \right. \tag{19}$$

After the integration of the time functions over the range $\left[0, \frac{2.\pi}{\omega}\right]$, one gets a non-linear eigen value problem, written in a matrix form as [8]:

$$2.\mathrm{K}.\vec{\mathrm{A}} + 3.\mathrm{B}\left(\vec{\mathrm{A}}\right).\vec{\mathrm{A}} = \omega^2.\mathrm{M}.\vec{\mathrm{A}} \tag{20}$$

$\vec{\mathrm{A}}$ is the column vector of the basic function contribution coefficients. K and M are the classical rigidity and mass matrices, well known in linear vibration theory, and $\mathrm{B}\left(\vec{\mathrm{A}}\right)$ is the nonlinear geometrical rigidity tensor. Equation (20) is the Benamar's adaptation of the Rayleigh-Ritz method to the nonlinear vibration problem, to be solved numerically, or explicitly. From Eq. (20), it is possible to calculate the frequency ω by pre-multiplying the two hand sides of the equation by $\vec{\mathrm{A}}^T$ which gives:

$$\omega^2 = \frac{\vec{\mathbf{A}}^{\mathbf{T}}.\mathbf{K}.\vec{\mathbf{A}} + \frac{3}{2}.\mathbf{A}^{\mathbf{T}}.\mathbf{B}\left(\vec{\mathbf{A}}\right).\vec{\mathbf{A}}}{\vec{\mathbf{A}}^{\mathbf{T}}\mathbf{M}.\vec{\mathbf{A}}} \tag{21}$$

The single mode approach (SMA), consists of neglecting all the basic functions except a single "resonant" mode. Thus, it reduces the multi-degree-of-freedom problem to a single dof. The single mode approach is often used in the literature [4] due to the great simplification it introduces in the theory on one hand, and on the other hand because the error it introduces in the estimation of the amplitude dependent nonlinear frequencies remains very small. Applying the SMA to Eq. (21) gives:

$$\omega^2 = \frac{\mathrm{K}_{11}}{\mathrm{M}_{11}} + \frac{3}{2}\frac{\mathrm{a}^2.\mathrm{B}_{1111}}{\mathrm{M}_{11}} \tag{22}$$

In which K_{11}, M_{11} and B_{1111} are the parameters related to the single mode examined, which is in the present case the fundamental mode of the plate shown in Fig. 1. Figure 3 shows, for a validation purpose, a satisfactory comparison between the results obtained here and those given in [4], corresponding to CSSCSS plate. Figure 4 gives

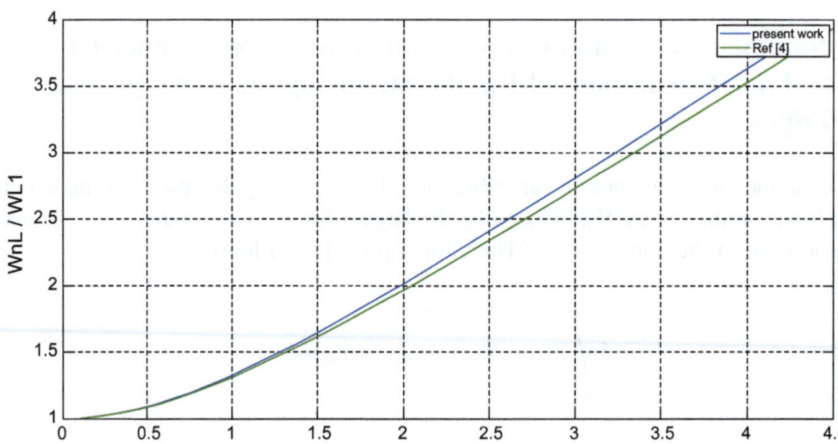

Fig. 3 Comparison between the present backbone curve and that of Ref. [4]. Aspect ratio = 0.66

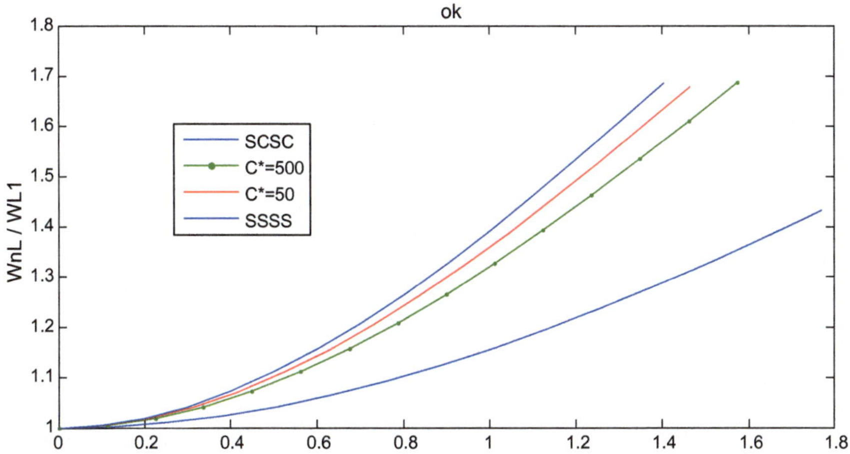

Fig. 4 Backbone curves for the four plates: (1) a CC plate in the x direction and SS in the y direction (2) a CC plate in the x direction and SS in the y direction with distributed rotational springs of stiffness $C^* = 50\left(C^* = \frac{\mathbf{K}_{iRota}.\mathbf{b}}{\mathbf{D}}\right)$; (3) a CC plate in the x direction and SS in the y direction with distributed rotational springs of stiffness C* = 500; (4) a CC plate in the x direction and SS in the y direction

the backbone curves corresponding to various values of the stiffness of the rotational springs distributed at the edges y = 0 and y = b of the rectangular plate. A hardening type non-linearity, indicating an increase in the frequency with the vibration amplitude is noticed and appears to become, as may be expected, more pronounced with increasing the spring stiffness.

5 Conclusion

To investigate the vibration of the plate shown in Fig. 1, the Rayleigh-Ritz method has been used in the linear case, with plate functions defined as products of x and y beam functions, with appropriate end supports in each direction. The extension of the Rayleigh-Ritz method to the nonlinear case, developed and applied to various non-linear problems by Benamar and his co-authors, has been used here to investigate the plate large vibration amplitudes. The basic functions used are obtained as product of beam functions in the x and y directions corresponding respectively to SS and ER (elastically restrained) beam end conditions, the last case being first analytically treated and numerically validated. Analytical details have been given and the numerical results were compared to those available in literature. The backbone curves are given for plates having different combinations of stiffness and aspect ratios.

References

1. El Bikri K, Benamar R, Bennouna M c (2003) Geometrically non-linear free vibrations of clamped simply supported rectangular plates. Part I: the effects of large vibration amplitudes on the fundamental mode shape. Comput Struct 81:2029–2043
2. El Kadiri M, Benamar R (2002) Improvement of the semi-analytical method, for determining the geometrically non-linear response of thin straight structures: part ii—first and second non-linear mode shapes of fully clamped rectangular plates. J Sound Vibr 257(1):19–62
3. Adri A, Beidouri Z, El Kadiri M, Benamar R Frequencies and mode shapes of a beam carrying a concentrated mass at different locations, with consideration of the effect of geometrical nonlinearity. An analytical approach and a parametric
4. Beidouri Z, Benamar R, El Kadiri M (2006). Geometrically non-linear transverse vibrations of C–S–S–S and C–S–C–S rectangular plates. Int J Non-Lin Mech 41:57–77
5. Formulas structural dynamics; IGORA A.KARNOVSKY- OLGA I.LEBED
6. Leissa AW (1969) Vibration of plates, NASA-SP-160. U.S.Government Printing Office, Washington
7. Li+ WL (2004) Vibration analysis of rectangular plates with general elastic boundary supports. J Sound Vibr 273:619–635
8. Benamar R, Bennouna MMK, White RG (1991) The effects of large vibration amplitudes

Preliminary Analysis of Temperature History When Milling Polymer Matrices for Fibre Reinforced Composites

F. Guesmi[1], A. Mkaddem[2], M. Beyaoui[1], A. Al-Zahrani[2], A. Jarraya[1,2(✉)], and M. Haddar[1]

[1] LA2MP, National Engineering School of Sfax (ENIS), University of Sfax, Road Soukra, 3038 Sfax, Tunisia
jarraya.abdessalem@gmail.com
[2] Department of Mechanical and Materials Engineering, Faculty of Engineering, University of Jeddah, PO Box 80327, Jeddah 21589, Saudi Arabia

Abstract. This paper addresses the temperature effects in abrasive milling of polymer matrices used in fibre reinforced polymers (FRP). Current knowledge of fibre reinforced polymers machining still remains understudied and involves challenging issues for their optimal use. A design of experiments including three matrix types was constructed to analyse the heat distribution due to cutting conditions. The cutting speed *N*, depth of cut a_p, and feed rate V_f were kept constant in this study. Milling tests were performed on resin specimens using an abrasive grinding wheel and 5-Axis CNC Machine with maximum spindle of 12,000 rpm. Three resinous matrices, namely, polyamide, Epoxy and Polyester were considered for the design of experiments. Equidistant thermocouples type K were embedded within the median plan of the specimen in order to record the temperature histories during cutting step. Measurements showed unequal temperature peaks on the three matrices referring to heat losses irrespective to matrix type. Temperature distribution within the specimens informs about heat evacuation during tool advance. The temperature histories reveal three typical regions regardless the matrix type being milled: fast heating period, maximum peak, and a relatively longer cooling period giving rise to room temperature. Those three periods vary sensitively with the resin properties.

Keywords: Temperature · Milling · Epoxy · Polyamide · Polyester

1 Introduction

Composite materials are widely used in various sectors such as mechanics, automotive, aeronautics and the aerospace because of their high specific mechanical properties. However, working and workability of these materials, i.e. by material removal volume, still remain challenging issues in several applications. Particularly, machining of these materials reveals damage in different type of composite structures.

The material structures are commonly pointed out as a main obstacle to control efficiently the material removal process [1–4]. In particular, heat generation and distribution to be involved due to machining should be controlled throughout composite-

F. Chaari et al. (eds.), *Advances in Materials, Mechanics and Manufacturing*, Lecture Notes in Mechanical Engineering,
https://doi.org/10.1007/978-3-030-24247-3_20

made components. Since matrix material is the continuous phase acting to bind the fiber reinforcement, and the fine interphase region i.e. the interface, its behavior plays a key role under thermomechanical loadings. The influence of temperature raise at the tool-tip could be reduced by selecting appropriate machining conditions in order to enhance tool lifetime [5, 6]. In spite the achievements within the last years [7], machinability of FRP still remain challenging. The physical phenomena to be involved during cutting controls sensitively the behavior at the tool-material interface [8–12]. Yashiro et al. [13] records the temperature history at the cutting point of CFRP composite. Machining generates high energy input per unit of material volume. This energy is converted into heat, which yields temperatures localisation and thermal damage into the material. During grinding [2], the thermocouple junction is exposed and bonded to the workpiece by smearing of the workpiece material, thereby providing direct contact with the material and, hence, ensure direct measurement of temperature at that point. Experimental studies were also conducted to analyse the temperature history in dry milling [12, 13]. The dynamic response of the measurement system is important in order to measure the temperature change in high-speed intermittent cutting with an end-mill. If the cutting temperature exceeds the glass-transition temperature of the matrix resin, it generally causes delamination and reduces the material strength.

In FRP, the matrix can be of two types: thermosetting or thermoplastic. Thermosetting resins are the most widespread in recent years for many applications. Two main resins are commonly used for manufacturing FRP composites, namely, polyesters and epoxy. These matrix resins are initially in viscous liquid state. The main advantages of these resins results in their ease of implementation, their rapid cross-linking without elimination of secondary products and their good mechanical properties. Epoxy resins are generally used for high performance applications while thermoplastic matrix composites become increasingly important in certain sectors such as aeronautics. Polyamides have also good mechanical properties and good thermal behavior under melting point. It is however worth noting that all aforementioned polymeric matrices are very sensitive to temperature variation [14]. Reyne [15], among others, have proposed the analysis of the resins behavior during cutting. Some properties of matrices have direct influences on their machinability.

This paper addresses the temperature analysis during milling of three resinous matrices, namely, epoxy, polyester, and polyamide. Milling tests were conducted using abrasive wheel of 10 mm in diameter and 8 mm in height. Peak values and temperature distribution obtained were specially discussed.

2 Experimental Setup and Method

Cutting tests were performed on epoxy resin parts using an abrasive diamond-grinding wheel. The grinding wheel (Diamond grinding points—galvanic metal bonding—reference D126) used in milling tests was supplied by GARANT Co. It is of diameter 10 mm and a head length of 8 mm. The grinding wheel with selected because of excellent cutting quality and optimum service life ensured by the galvanic nickel binder. Many diamond particles are bunched or glued to the rod or tool body and act as multiple cutting points. The matrix used was provided by UNIVERS RESINE

Company and results in a mixture of an epoxy-type thermosetting resin. Its implementation is carried out by infusion at room temperature.

Table 1 summarizes the material properties of the three resins used. The panels are melded in a mold of size 300 × 250 × 14 mm^3, using a mixture of 75% resin and 25% hardener. The panels were then laser cut to specimens of dimension 80 × 40 × 14 mm^3 each. Subsequently, the specimens obtained were face milled to reach the thickness of 8 mm, on a universal milling machine model MOMAC FU-1S of 1800 rpm maximum speed, and 7 kW power.

Table 1 Proprieties of the matrices considered in the design of experiments

Matrix type	Property	Value
Epoxy	Density ρ Thermal conductivity at 20 °C Dry extract in volume	1.27 ± 0.05 g/cm^3 0.2 W/m × °C 33 ± 2%
Polyester	Density ρ Thermal conductivity at 20 °C Dry extract in volume	1.1 ± 0.05 g/cm^3 0.2 W/m × °C 31 ± 1%
Polyamide	Density ρ Thermal conductivity at 20 °C Dry extract in volume	1.14 ± 0.05 g/cm^3 0.25 W/m × °C 33 ± 2%

The temperature is recorded in the median plan through the thickness of the specimen, using four thermocouples TCs (TC1, TC2, TC3 and TC4) type K, supplied by ELECTRONIC SHOP Co. Their maximum measuring capacity, See beck coefficient(α) and standard deviation are −270 to +1370 °C/−454 to +2498 °F, 39.45 μV/°C at 0 °C and 2.2–0.75%, respectively. The TCs are placed on the test specimen 20 mm from each other inside a pre-drilled holes of 2 mm diameter, at a distance of 0.5 mm away the trim plan. The test specimen equipped with the TCs is illustrated in Fig. 1.

The tests were carried out on a Spinner U-620 5-Axis CNC Machine with maximum spindle speed of 12,000 rpm and maximum power of 19 kW. The experimental set-up was shown in Fig. 2.

The experimental device was connected to an acquisition system including computer unit with an adapted LabVIEW program able to recording temperature history outputs over the four TCs simultaneously at 0.1 s^{-1} while the tool advances through the specimen.

The tool characteristics are shown in Table 2.

The profile milling operation (trimming) is carried out without lubrication with a 10 mm diameter galvanic diamond-grinding wheel on specimens of dimensions (80 × 40 × 8) mm^3. During grinding, thermocouples measure temperatures at a distance of 0.5 mm below the surface of the workpiece. All the milling operations were carried out under the dry cutting conditions, and each experiment was repeated two times. The parameters used, and their ranges are presented in Table 3.

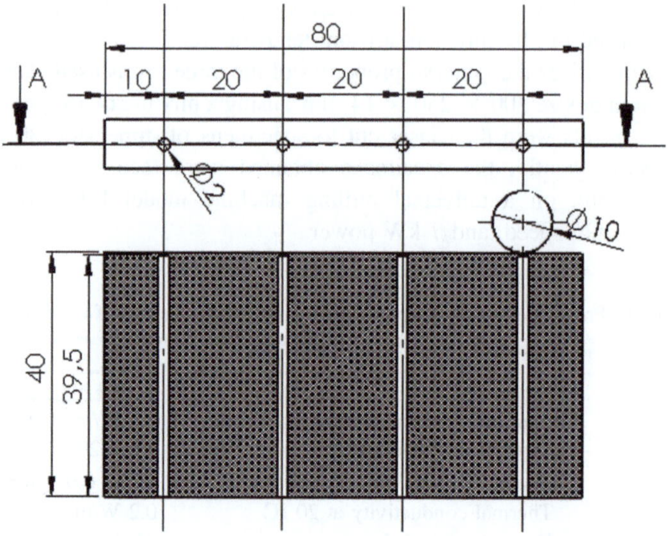

Fig. 1 Specimen design, TCs and tool location during cutting test of polymer matrix

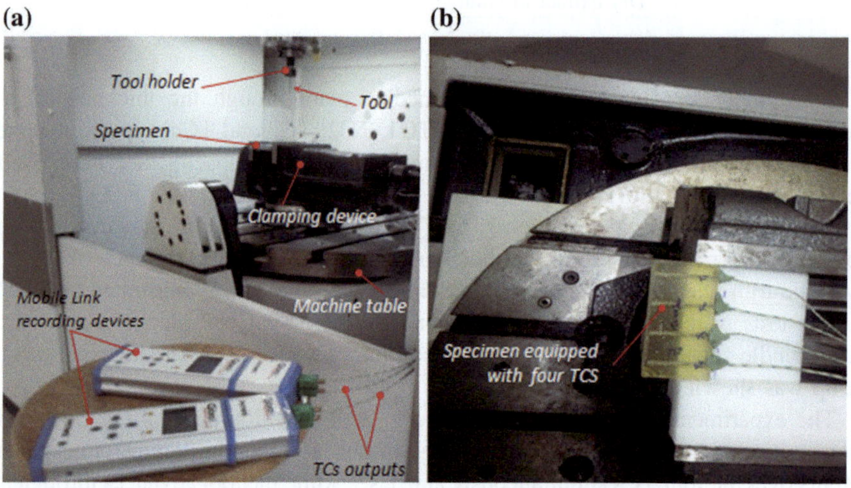

Fig. 2 Experimental set-up. **a** Details of the testing device, and **b** enlarged view of the specimen equipped with the TCs

3 Results and Discussion

The results of Fig. 3a show typical plots temperature histories recorded for polyamide matrix using 4 thermocouples. It may be noticed an overall tendency for all curves presenting, in turn, temperature fall after reaching a specified maximum in front of the

Table 2 Tool parameters

Abrasives	Diamond
Abbreviation of the abrasive	D
Diameter tail	6 mm
Head diameter	10 mm
Head length	8 mm
Form	Cylindrical
Total length	60 mm

Table 3 Comparison of the frequency parameters $\frac{\omega^2.a^2.b^2.\rho}{\pi^4.D}$ (b) calculated here with (a) results given in [8] for different values of aspect ratio and different values of stiffness

	$\alpha = 0,4$		$\alpha = 0,6$		$\alpha = 0,8$		$\alpha = 1$	
	(a)	(b)	(a)	(b)	(a)	(b)	(a)	(b)
0	8,41	8,41	5,138	5,1388	4,203	4,2035	4,000	4,000
0, 5	8,438	8,4415	5,206	5,2081	4,326	4,3271	4,194	4,1941
1	8,468	8,4713	5,271	5,2736	4,441	4,4423	4,373	4,373
2	8,525	8,5261	5,387	5,3919	4,646	4,6483	4,691	4,691
3	8,572	8,5754	5,491	5,4958	4,825	4,8271	4,965	4,966
4	8,611	8,6199	5,584	5,5877	4,982	4,9837	5,205	5,2052
5	8,649	8,6604	5,666	5,6696	5,120	5,1219	5,415	5,4152
10	8,801	8,8169	5,974	5,9744	5,625	5,6252	6,169	6,1693
50	9,204	9,2167	6,671	6,672	6,719	6,723	7,777	7,7833
100	9,312	9,327	6,845	6,8452	6,981	6,9835	8,137	8,144
500	-	9,4382	6,974	7,0168	7,234	7,2382	8,494	8,505
∞	9,448	9,47	7,059	7,065	7,304	7,3091	8,593	8,605

TCs, respectively. The findings refer physically to heating and cooling period during tool passage. Cooling step mark much higher duration to reach room temperature than heating step. The curves exhibit similar three typical regions:

(i) Quasi-static evolution reflecting stable temperature that precedes the starting of the milling operation.

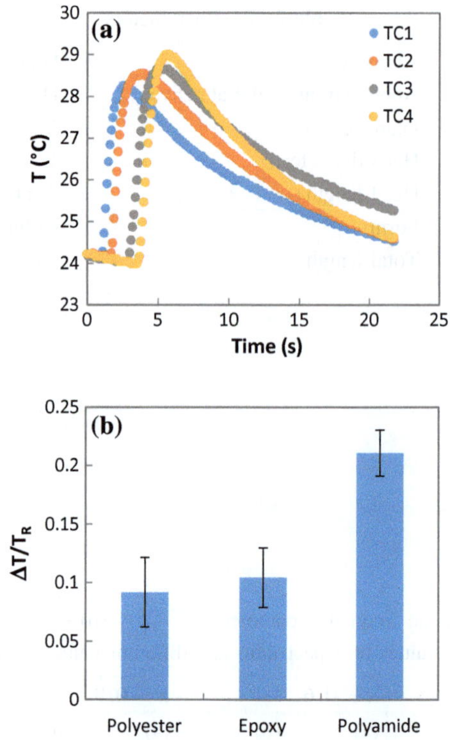

Fig. 3 Temperature histories versus time. **a** Typical temperature histories recorded for polyamide, and **b** relative peak average temperature for the three considered matrices

(ii) Sudden temperature raise resulting in the approach of the tool face to the thermocouple level. The curves reach their maximum peaks at the passage of the tool in front of the thermocouples, respectively.

(iii) Slow temperature drop over approximately 75% of the total milling period marking the move of the tool away from the TCs locations, respectively. This should draw a cooling step that marks much more time than heating previously reached.

Figure 3b shows the relative peak average of temperature obtained at the aforementioned milling conditions. Polyamide labels the highest peak average resulting in approximately a double value compared with the values obtained for polyester and epoxy matrices while the peak averages of the two last matrices can be roughly confounded.

4 Conclusions

In this attempt, an experimental approach is proposed to investigate the thermal behavior of composite matrices under milling. Temperature histories locally recorded at four milling stages were examined to explain heat distribution within the polymeric matrices. It appears that the cutting temperature increases rapidly as the tool approaches the TCs, marking a peak just in front of TCs respectively, and then, falls much more slowly up to reaching room temperature.

Three typical regions were detected in the temperature curve regardless the matrix type being milled: heating, maximum peak, and slight cooling up to room temperature.

References

1. Sheikh-Ahmad JY (2009) Machining of polymer composite. Springer, New York, p 210
2. Chockalingam P, Kuang KC, Vijayaram TR (2013) Effects of grinding process parameters and coolants on the grindability of GFRP laminates. Mater Manuf Process 28:1071–1076
3. Mkaddem A, Demirci I, El Mansori M (2008) A micro–macro combined approach using FEM for modelling of machining of FRP composites: cutting forces analysis. Compos Sci Technol 68:3123–3127
4. Ben Soussia A, Mkaddem A, El Mansori M (2014) Rigorous treatment of dry cutting of FRP-interface consumption concept: a review. Int J Mech Sci 83:1–29
5. Hanasaki S, Fujiwara J, Nomura M (1994) Tool wear mechanism in cutting of CFRP. Trans Jpn Soc Mech Eng Ser C 60(569):297–302
6. Hanasaki S, Fujiwara J, Kawai T, Nomura M, Miyamoto T (2005) Study on tool wear mechanism of CFRP cutting II. Trans Jpn Soc Mech Eng Ser C 71(702):719–724
7. Arola D, Sultan MB, Ramulu M (2002) Finite element modeling of edge trimming fiber reinforced plastics. J Manuf Sci Eng-Trans ASME 124:32–41
8. Grzesik W, Nieslony P (2004) Physics based modelling of interface temperatures in machining with multilayer coated tools at moderate cutting speeds. Int J Mach Tool Manuf 44:889–901
9. Mkaddem A, Ben Soussia A, El Mansori M (2013) Wear resistance of CVD and PVD multilayer coatings when dry cutting fiber reinforced polymers (FRP). Wear 302:946–954
10. Brinksmeier E, Fangmann S, Rentsch R (2011) Drilling of composites and resulting surface integrity. CIRP Ann Manuf Technol 60:57–60
11. Tian N, Zhou (2011) A sensitivity and uncertainty analyses for ignition of fiber-reinforced polymer panels. Fire Mater 1:120
12. Liu J, Chen G, Ji C, Qin X, Li H, Ren C (2014) An investigation of workpiece temperature variation of helical milling for carbon fiber reinforced plastics (CFRP). Int J Mach Tool Manuf 86:89–103
13. Yashiro T, Ogawa T, Sasahara H (2013) Temperature measurement of cutting tool and machined surface layer in milling of CFRP. Int J Mach Tools Manuf 70:63–69
14. Botelho EC, Rezende MC (2006) Monitoring of Carbon fiber/polyamide composite processing by rheological an thermal analyses. Polym Plast Technol Eng 45:61–69
15. Reyne M (1998) Technologie des composites, 3th ed. Hermes editions. ISBN 2- 86601:688-8

Prediction of the Ductility Limit of Magnesium AZ31B Alloy

Mohamed Yassine Jedidi[1,2(✉)], Mohamed Ben Bettaieb[2,3], Anas Bouguecha[1], Farid Abed-Meraim[2,3], Mohamed Taoufik Khabou[1], and Mohamed Haddar[1]

[1] Laboratoire de Mécanique, Modélisation et de Production (LA2MP), École Nationale d'Ingénieurs de Sfax (ENIS), Route Soukra Km 3.5, Sfax, Tunisia
mohamed-yasine.jdidi@enis.tn, anas.bouguecha@gmx.de, mtkhabou@hotmail.com, mohamed.haddar@enis.rnu.tn

[2] Arts et Métiers ParisTech, Université de Lorraine, CNRS, LEM3, 57000 Nancy, France
{mohamed.benbettaieb, farid.abed-meraim}@ensam.eu

[3] DAMAS, Laboratory of Excellence on Design of Alloy Metals for low-mass Structures, Université de Lorraine, Nancy, France

Abstract. In many engineering applications (automotive, computer and mobile device industries, etc.), magnesium alloys have been widely used owing to their interesting physical and mechanical parameters. However, magnesium alloys are identified by the low ductility at room temperature, due to their strong plastic anisotropy and the yielding asymmetry between tension and compression. In this work, the ductility limit of a rolled magnesium AZ31 sheet metal at room temperature is numerically investigated. This investigation is based on the coupling between a reduced-order crystal plasticity model and the Marciniak–Kuczyński localized necking approach. This reduced-order model is used to describe the anisotropic behavior of this material taking into account the strong plastic anisotropy (e.g., yielding asymmetry between tension and compression) due to the limited number of slip systems (i.e., twinning mode). To accurately describe the plastic anisotropy due to slip and twinning modes, a combination of two separate yield functions (according to Barlat and Cazacu) is used. The coupling between the adopted constitutive framework and the Marciniak–Kuczyński instability approach is numerically implemented via an implicit algorithm. Comparisons between experimental results from the literature and numerical results obtained by using our calculation tool are carried out to validate the choice of the reduced-order crystal plasticity model.

Keywords: Ductility limit · Reduced-order crystal plasticity model · Marciniak–Kuczyński instability approach · Plastic anisotropy · Yielding asymmetry

F. Chaari et al. (eds.), *Advances in Materials, Mechanics and Manufacturing*, Lecture Notes in Mechanical Engineering,
https://doi.org/10.1007/978-3-030-24247-3_21

1 Introduction

The magnesium alloy AZ31B has been used in a wide range of engineering applications, notably in the aeronautic and automotive industries [17]. Compared to other materials with hexagonal structures, the magnesium alloy AZ31B is characterized by its lightweight and high specific strength. However, it is known to have low formability at room temperature, due to its strong plastic anisotropy and the yielding asymmetry between tension and compression. These characteristics are respectively caused by the limited number of slip systems and the activation of the twinning deformation mode. To accurately predict the plastic response of the magnesium alloy AZ31B, several researchers have developed constitutive models that take into account both deformation modes (namely slip and twinning). In this work, we have chosen a reduced-order crystal plasticity model. This model has been recently developed for magnesium alloys [8, 9, 13]. It allows describing the strong plastic anisotropy and its evolution as well as the yielding asymmetry between tension and compression, which results from the competition between slip and twin mechanisms. Hence, the plastic anisotropy is described in this reduced-order model by two separate yield functions: the Cazacu yield function [3] to describe the plastic anisotropy due to the activation of twinning mode and the Barlat yield function [2] to model slip-induced plastic anisotropy.

Plastic deformation during sheet metal forming processes is limited by the occurrence of localized necking. To predict the ductility limit, the concept of forming limit diagram (FLD) is introduced and applied in the present investigation. The determination of the ductility limit by experimental methods is often complex and relatively expensive. To avoid these practical difficulties, several theoretical approaches for localized necking analysis have been developed by different researchers to predict FLDs, such as: the Swift model [14], the Hill model [6], the Marciniak–Kuczyński (M–K) imperfection model [10], the bifurcation theory [12], the stability analysis by linear perturbation theory [4] and the modified maximum force criterion [1].

The M–K formability limit analysis is one of the most widely-used ductility approaches in the literature. Several researchers have coupled this approach with various constitutive models to predict forming limit diagrams of HCP materials. In this field, one may quote [16] who have coupled this approach with an elaborate version of the Cazacu model [11] to predict the FLD for magnesium AZ31B alloy.

It is well known that FLDs are strongly influenced by the mechanical properties of the used sheet metal, such as plastic anisotropy and hardening parameters. Hence, two hardening laws are applied with the combined yield function (Cazacu–Barlat).

In the current study, a generic Mathematica script is developed to couple the M–K approach with the reduced-order crystal plasticity model. The numerical predictions are compared with experimental results of [8] to validate the implementation of this reduced-order model. Then, the ductility limit of the AZ31B magnesium alloy is analyzed. Furthermore, the effect of the activated deformation mode on the ductility limit is also investigated.

Notations, conventions and abbreviations

Vectors and tensors are indicated by **bold** letters and symbols. Scalar variables and parameters are designated by thin letters or symbols.

The used Void convention leads to express all second-order symmetric tensors in vector form as: $\mathbf{V} = \left(\bullet_{11},\ \bullet_{22},\ \bullet_{33},\ \sqrt{2}\bullet_{12},\ \sqrt{2}\bullet_{23},\ \sqrt{2}\bullet_{13}\right)^{\mathrm{T}}$.

2 Reduced-Order Crystal Plasticity Model

The constitutive model proposed in [8] is adopted here to describe the material behavior of magnesium sheets. This model is formulated within the framework of rate-independent finite plasticity. As this model is used to predict the ductility limit of thin sheets, the plane-stress assumption is adopted (see, e.g., [7]).

To describe the anisotropic plasticity due to slip and twinning modes, two independent yield surfaces are introduced.

2.1 Twinning Mode

Twinning leads to yielding asymmetry between tension and compression. For this reason, the non-quadratic yield criterion developed by Cazacu et al. [3] is used. Within this criterion, the effective stress $\tilde{\sigma}^t$ for the twinning mode is expressed as:

$$\tilde{\sigma}^t\left(\sum_1^t, \sum_2^t, \sum_3^t\right) = \left(\left(\left|\sum_1^t\right| - k\sum_1^t\right)^{a_t} + \left(\left|\sum_2^t\right| - k\sum_2^t\right)^{a_t} + \left(\left|\sum_3^t\right| - k\sum_3^t\right)^{a_t}\right)^{1/a_t}, \tag{1}$$

where $\sum_1^t$, $\sum_2^t$ and $\sum_3^t$ are the eigenvalues of tensor $\sum^t$ and k (comprised between −1 and 1) and a_t (superior to 1) are material parameters describing the initial stress differential effect and the shape of the yield surface, respectively. Tensor $\sum^t$ is obtained by a linear transformation from the deviatoric Cauchy stress tensor **s**; $\sum^t = \mathbf{L}^t : \mathbf{s}$. The transformation tensor $\mathbf{L}^t$ is given in the following matrix form:

$$\mathbf{L}^t = \begin{pmatrix} l_{LL}^t & L_{LT}^t & L_{LS}^t & 0 & 0 & 0 \\ L_{LT}^t & l_{TT}^t & L_{TS}^t & 0 & 0 & 0 \\ L_{LS}^t & L_{TS}^t & l_{SS}^t & 0 & 0 & 0 \\ 0 & 0 & 0 & l_{LT}^t & 0 & 0 \\ 0 & 0 & 0 & 0 & l_{TS}^t & 0 \\ 0 & 0 & 0 & 0 & 0 & l_{SL}^t \end{pmatrix}, \tag{2}$$

where $\mathbf{L}^t$ is expressed onto three orthotropic axes using three directions: longitudinal (L), transverse (T) and short-transverse (S).

The activation of the twinning deformation mode is governed by the following Kuhn–Tucker inequality constraints:

$$\phi^t = \tilde{\sigma}^t - Y^t \leq 0; \quad \dot{\tilde{\varepsilon}}^t \geq 0; \quad \phi^t \dot{\tilde{\varepsilon}}^t = 0 \tag{3}$$

where Y^t is the hardening yield function and $\dot{\tilde{\varepsilon}}^t$ is the equivalent plastic strain rate due to twinning. Y^t is expressed as a function of the cumulated plastic strain due to slip $\tilde{\varepsilon}^s$ and twinning $\tilde{\varepsilon}^t$:

$$Y^t\left(\tilde{\varepsilon}^s, \tilde{\varepsilon}^t\right) = R_0^t + Q_1^t\left(1 - e^{-b_1^t \tilde{\varepsilon}^t}\right) + Q^t\left(e^{q^t \tilde{\varepsilon}^t} - 1\right) + H^t \tilde{\varepsilon}^t + H^{st} \tilde{\varepsilon}^s. \tag{4}$$

The plastic strain rate $\dot{\varepsilon}^t$ due to twinning is given by the normality law:

$$\dot{\varepsilon}^t = \dot{\tilde{\varepsilon}}^t \frac{\partial \phi^t}{\partial \boldsymbol{\sigma}}(\boldsymbol{\sigma}). \tag{5}$$

The equivalence relationship of the plastic work rate allows us to determine the expression of $\dot{\tilde{\varepsilon}}^t$:

$$\dot{\tilde{\varepsilon}} = \frac{\boldsymbol{\sigma} : \dot{\boldsymbol{\varepsilon}}^t}{\tilde{\sigma}^t}. \tag{6}$$

2.2 Slip Mode

Limited number of slip systems for HCP materials leads to a strong initial and evolving plastic anisotropy. For this reason, the yield criterion proposed by Eyckens et al. [2] and based on a linear transformation technique is used in this work. Within this criterion, the effective stress $\tilde{\sigma}^s$ for the slip mode is expressed as:

$$\tilde{\sigma}^s\left(\sum_1^s, \sum_2^s, \sum_3^s\right) = \left(\frac{1}{2}\left(\left|\sum_1^s - \sum_2^s\right|^{a_s} + \left|\sum_2^s - \sum_3^s\right|^{a_s} + \left|\sum_3^s - \sum_1^s\right|^{a_s}\right)\right)^{1/a_s}, \tag{7}$$

where $\sum_1^s$, $\sum_2^s$ and $\sum_3^s$ are the eigenvalues of tensor $\sum^s$ and a_s (superior to 1) is a material parameter describing the shape of the yield surface. Tensor $\sum^s$ is obtained by a linear transformation from the Cauchy stress tensor: $\sum^s = \mathbf{L}^s : \boldsymbol{\sigma}$. The transformation tensor $\mathbf{L}^s$ is given in the following matrix form:

$$\mathbf{L}^s = \begin{pmatrix} \frac{1}{3}\left(L_{TT}^s + L_{SS}^s\right) & -\frac{1}{3}L_{SS}^s & -\frac{1}{3}L_{TT}^s & 0 & 0 & 0 \\ -\frac{1}{3}L_{SS}^s & \frac{1}{3}\left(L_{SS}^s + L_{LL}^s\right) & -\frac{1}{3}L_{LL}^s & 0 & 0 & 0 \\ -\frac{1}{3}L_{TT}^s & -\frac{1}{3}L_{LL}^s & \frac{1}{3}\left(L_{TT}^s + L_{LL}^s\right) & 0 & 0 & 0 \\ 0 & 0 & 0 & L_{TL}^s & 0 & 0 \\ 0 & 0 & 0 & 0 & L_{TS}^s & 0 \\ 0 & 0 & 0 & 0 & 0 & L_{SL}^s \end{pmatrix}. \tag{8}$$

The activation of the slip deformation mode is governed by the following Kuhn–Tucker inequality constraints:

$$\phi^{s} = \tilde{\sigma}^{s} - Y^{s} \leq 0; \quad \dot{\tilde{\varepsilon}}^{s} \geq 0; \quad \phi^{s}\dot{\tilde{\varepsilon}}^{s} = 0, \tag{9}$$

where Y^s is the hardening yield function and $\dot{\tilde{\varepsilon}}^{s}$ is the equivalent plastic strain rate due to slip. Y^s is expressed as a function of the cumulated plastic strain due to slip $\dot{\tilde{\varepsilon}}^{s}$ and twinning $\dot{\tilde{\varepsilon}}^{t}$:

$$Y^{s}(\tilde{\varepsilon}^{s}, \tilde{\varepsilon}^{t}) = R_{0}^{s} + Q_{1}^{s}\left(1 - e^{-b_{1}^{s}\tilde{\varepsilon}^{s}}\right) + Q_{2}^{s}\left(1 - e^{-b_{2}^{s}\tilde{\varepsilon}^{s}}\right) + H^{ts}\tilde{\varepsilon}^{t}. \tag{10}$$

The plastic strain rate $\dot{\varepsilon}^{s}$ is given by the normality law:

$$\dot{\varepsilon}^{s} = \dot{\tilde{\varepsilon}}^{s}\frac{\partial \phi^{s}}{\partial \boldsymbol{\sigma}}(\boldsymbol{\sigma}). \tag{11}$$

The equivalence relationship of the plastic work rate allows us to determine the expression of $\dot{\tilde{\varepsilon}}$:

$$\dot{\tilde{\varepsilon}}^{t} = \frac{\boldsymbol{\sigma} : \dot{\boldsymbol{\varepsilon}}^{s}}{\tilde{\sigma}^{s}}. \tag{12}$$

2.3 Combined Yield Surface

The plastic anisotropy of the studied HCP materials is defined by a combination of the Cazacu and Barlat yield functions as shown in Fig. 1.

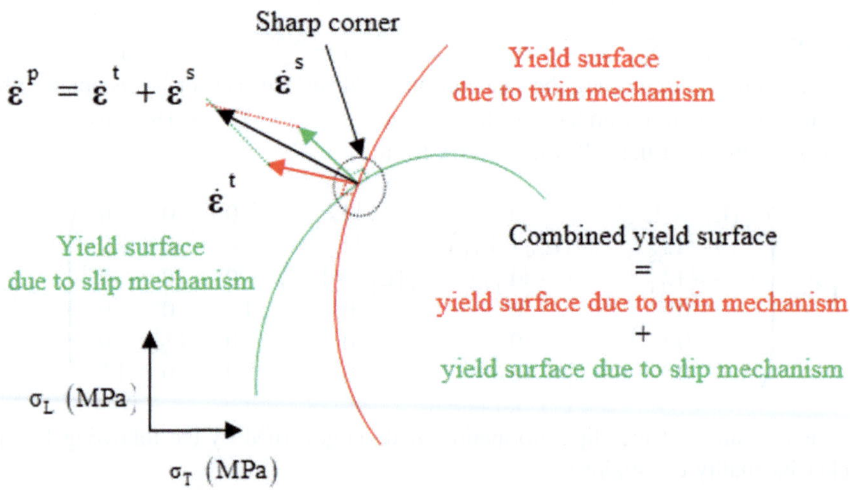

Fig. 1 Schematic illustration of the combined yield surface

The plastic strain rate given by Eq. (5) [resp. Eq. (11)] represents only the contribution to the total plastic strain rate $\dot{\varepsilon}^{\mathrm{p}}$ when the twinning mode (resp. the slip mode) is activated. When both deformation modes are activated simultaneously, the stress state is located at the intersection of the two yield surfaces as illustrated in Fig. 1, and the plastic strain rate $\underline{\dot{\varepsilon}}^{\mathrm{p}}$ is expressed as:

$$\dot{\varepsilon}^{\mathrm{p}} = \dot{\varepsilon}^{\mathrm{t}} \frac{\partial \phi^{\mathrm{t}}}{\partial \boldsymbol{\sigma}}(\boldsymbol{\sigma}) + \dot{\varepsilon}^{\mathrm{s}} \frac{\partial \phi^{\mathrm{s}}}{\partial \boldsymbol{\sigma}}(\boldsymbol{\sigma}). \tag{13}$$

3 M–K Localized Necking Criterion

The theoretical M–K approach assumes that the sheet is composed of two zones: the homogeneous zone (a) and the band zone (b). Figure 2 illustrates this concept, where the band is characterized by its orientation θ with respect to the major strain direction, and its relative thickness f with respect to the homogeneous zone ($\mathrm{f} = \mathrm{t}^{\mathrm{b}}/\mathrm{t}^{\mathrm{a}}$). The relative thickness f is also called the imperfection factor or imperfection ratio, and its initial value classically ranges between 0.98 and 0.999 [5].

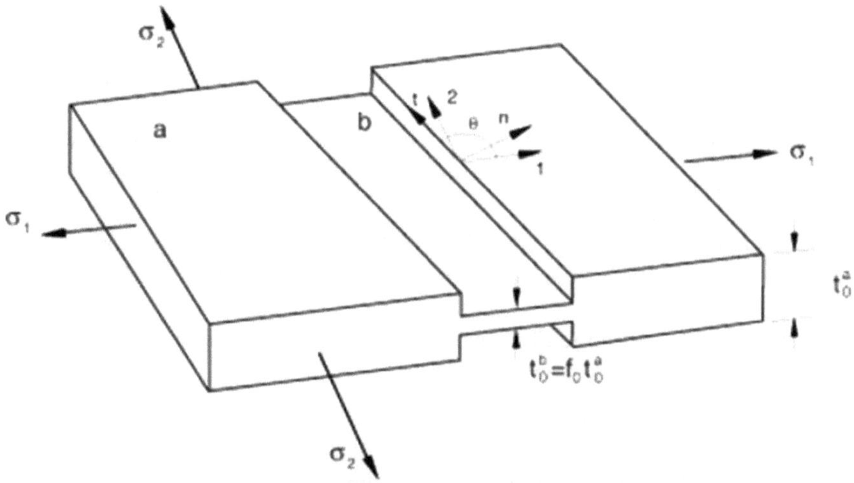

Fig. 2 Marciniak–Kuczyński (M–K) model

The current band orientation θ is related to its initial counterpart θ_o by the following relation:

$$\mathrm{tang}(\theta) = \mathrm{tang}(\theta_o)\mathrm{e}^{\left(\varepsilon_{11}^{\mathrm{a}} - \varepsilon_{22}^{\mathrm{a}}\right)}, \tag{14}$$

where $\varepsilon_{11}^{\mathrm{a}}$ and $\varepsilon_{22}^{\mathrm{a}}$ are the major and the minor strain in the plane of the sheet (in the homogenous zone).

In addition to the constitutive equations described in Sect. 2, the M–K approach is defined by the following equations:

- The equation describing the force equilibrium between the band and the homogeneous zone:

$$\dot{\sigma}^{a}.\vec{\mathbf{n}}t^{a} = \dot{\sigma}^{b}.\vec{\mathbf{n}}t^{b}. \tag{15}$$

- The geometric compatibility condition linking the velocity gradient in the homogeneous zone $\mathbf{L}^{a}$ to its counterpart in the band zone $\mathbf{L}^{b}$: allowing us to relate the strain rate fields in the two zones:

$$\mathbf{L}^{b} = \mathbf{L}^{a} + \dot{\vec{\beta}} \otimes \vec{\mathbf{n}}, \tag{16}$$

where $\dot{\vec{\beta}}$ is the jump vector.

To predict the occurrence of strain localization, a biaxial loading is applied to the sheet metal, which is defined by the following velocity gradient in the homogeneous zone:

$$\mathbf{L}^{a} = \begin{bmatrix} L_{11}^{a} & 0 \\ 0 & \rho\, L_{11}^{a} \end{bmatrix}, \tag{17}$$

where ρ is the strain-path ratio. Due to the presence of the geometrical imperfection, the plastic strain will become more and more concentrated within the band (as compared to the homogeneous zone). Strain localization is assumed to occur when the strain-rate ratio $\left|\dot{\varepsilon}_{33}^{b}\right| / \left|\dot{\varepsilon}_{33}^{a}\right|$ reaches a critical value (this critical value is typically set to 10).

When this strain localization condition is satisfied, the current strain component ε_{11}^{a} is referred to as the critical strain ε_{11}^{*} corresponding to the strain-path ratio ρ and to the initial band orientation θ_0.

The prediction of the entire FLD using the M–K approach is based on the following incremental algorithm with two nested loops:

- For each strain-path ratio ranging from $\rho = -0.5$ to $\rho = 1$ (we take intervals of 0.1).
- For each initial band orientation ranging from $\theta_0 = 0°$ to $\theta_0 = 90°$ (with typical intervals of 1°).

Apply the implicit incremental algorithm based on the theoretical framework and constitutive equations presented above to calculate the critical strain ε_{11}^{*}.

The smallest critical strain ε_{11}^{*} over all possible initial band orientations θ_0 defines the localization limit strain ε_{11}^{L} corresponding to the strain-path ratio ρ. This major limit strain is used to calculate the associated minor limit strain ε_{22}^{L} using the corresponding strain-path ratio ρ.

4 Results and Discussions

The numerical simulations in the current work are based on the parameters for the AZ31 magnesium alloy, as shown in Table 1. For elastic behavior, this material is characterized by Young's modulus E = 43 GPa and Poisson's ratio $\nu = 0.29$.

Table 1 Isotropic hardening and plastic anisotropy parameters using slip and twin model for magnesium AZ31B alloy [8]

Slip	Hardening	$R_0^s = 206\,\text{MPa}$	$Q_1^s = 126\,\text{MPa}$	$H^{ts} = 637\,\text{MPa}$	
		$Q_2^s = 40.3\,\text{MPa}$	$b_2^s = 230$	$b_1^s = 23.6$	
	Anisotropy	$L_{LL}^s = 1$	$L_{TT}^s = 1.09$	$L_{SS}^s = 1.25$	$a_s = 4$
		$L_{LT}^s = 1.22$	$L_{TS}^s = 1.69$	$L_{SL}^s = 1.55$	
Twin	Hardening	$R_0^t = 30\,\text{MPa}$	$Q_1^t = 0\,\text{MPa}$	$b_1^t(\sim)$	$H^{st} = 386\,\text{MPa}$
		$Q^t = 19.4\,\text{MPa}$	$q^t = 6.3$	$H^t = 0$	
	Anisotropy	$l_{LL}^t = 1$	$l_{TT}^t = 0.97$	$l_{SS}^t = 0.33$	$a_t = 4$
		$l_{LT}^t = 0.78$	$l_{TS}^t = 0.61$	$l_{SL}^t = 0.60$	$k = 0.53$
		$L_{LT}^t = 0.22$	$L_{TS}^t = 0.76$	$L_{LS}^t = 0.73$	

In the current paper, the amount of deformation due to slip and twinning modes is predicted by numerical simulations. In the literature, this amount can be experimentally investigated by using the electron backscatter diffraction (EBSD) which is capable to observe the activation of different twin and slip systems. Based on EBSD, twin mode is identified from local intergranular misorientation maps, and active slip system is identified from long range intragranular misorientation maps [15]. Another diffraction type denoted X ray diffraction (XRD) is used in the literature to calculate the amount of deformation, but for materials with high symmetry. Thus, XRD is not valid for HCP materials due to their yielding asymmetry between tension and compression.

In a preliminary stage, some relevant comparisons are carried out in the current work, using the parameters identified in [8], to assess and validate the numerical implementation of the constitutive framework. Then, predictions of forming limit diagrams (FLDs) are undertaken for magnesium AZ3B alloy.

Experimental results were carried out based on uniaxial tension and compression tests for the AZ31B magnesium alloy to predict the engineering stress F/S_0, where F is the applied force in the tensile and compression tests, and S_0 represents the nominal section of the sheet. Moreover, numerical simulations are carried out, taking into account the anisotropy parameters as well as the hardening function. Despite the complexity of the material behavior, Fig. 3 clearly shows that the reduced-order model is able to reproduce the experimental results with good agreement, owing to flow stresses and plastic anisotropy parameters.

Accordingly, the constitutive modeling of the AZ31B magnesium alloy is validated compared with experiments in [8]. The aim of this investigation is to predict FLDs for this material based on the M–K approach using the reduced-order model. For this

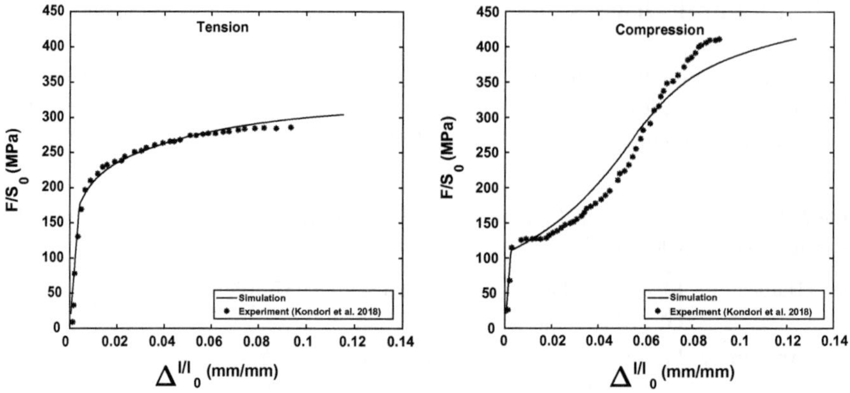

Fig. 3 Engineering stress versus axial strain

reason, the numerical simulation is carried out with a fixed imperfection factor f = 0.98.

The onset of localized necking is determined by the numerical necking criterion I_n (where I_n represents the strain-rate ratio $\left|\dot{\varepsilon}^{b}_{33}\right|/\left|\dot{\varepsilon}^{a}_{33}\right|$). More specifically, localized necking occurs when the strain-rate ratio I_n becomes larger than 10 for a given strain path and initial band orientation. Figure 4 shows the evolution of this localized necking indicator for four representative strain paths and an initial band orientation $\theta_0 = 0$: uniaxial tension ($\rho = -0.5$), plane-strain tension ($\rho = 0$), biaxial tension (e.g.; $\rho = 0.5$), and equibiaxial expansion ($\rho = 1$).

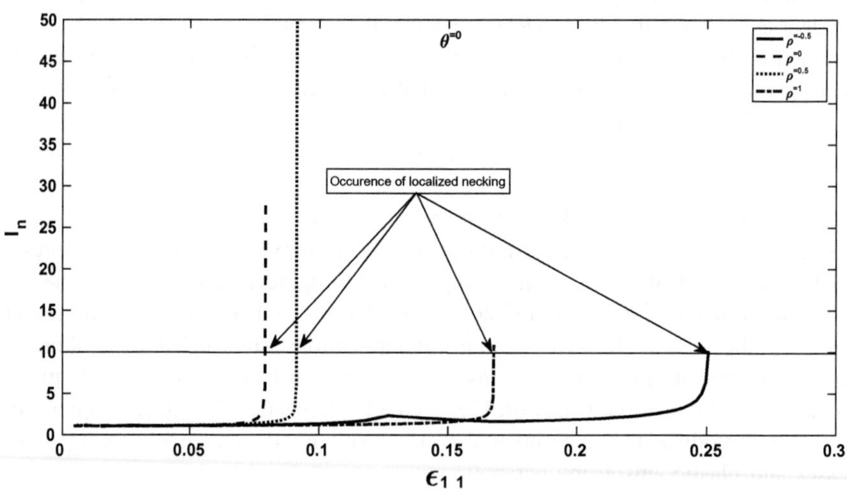

Fig. 4 Prediction of localized necking

The imperfection factor depends on thicknesses t^a, t^b and the initial imperfection factor f_0. It is always smaller than f_0 during sheet deformation (see Fig. 5).

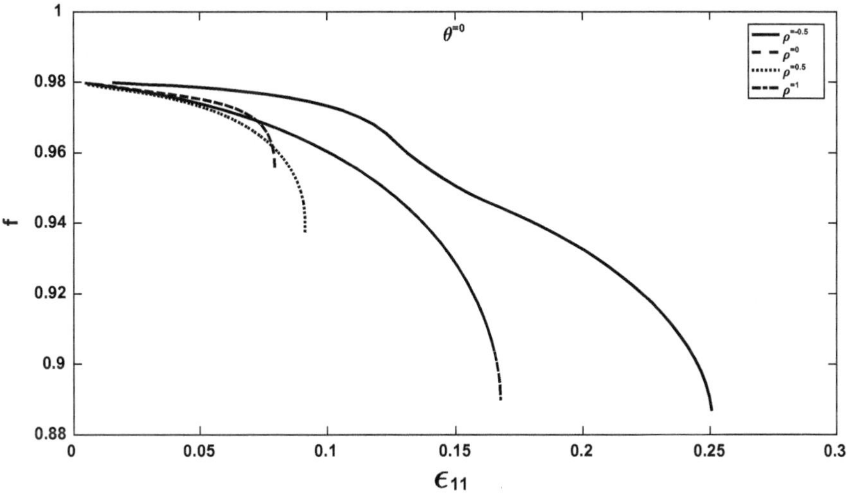

Fig. 5 Imperfection factor evolution

The necking criterion of Sect. 3 allows us to predict the FLD of the AZ31B magnesium sheet metal. Figure 6 compares the FLD predicted by the proposed reduced-order Cazacu–Barlat model to that obtained by the simpler Barlat–von Mises model.

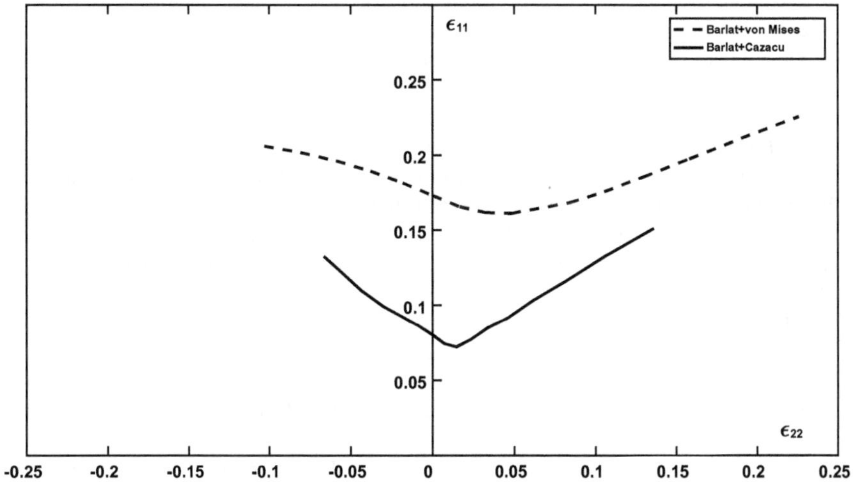

Fig. 6 Influence of the adopted yield criterion on the predicted FLD

To recover the classical von Mises criterion from the Cazacu criterion, the following parameters should be taken: $a_t = 2$, $k = 0$, and $\mathbf{L}^t = \mathbf{I}$, where $\mathbf{I}$ is the identity tensor.

The FLD predicted with the reduced model is limited to the strain between −0.07 and 0.15, whereas the FLD based on von Mises-Barlat model is ranged from −0.10 to 0.20. This is due to the limits of validity of the proposed model. From these predicted FLDs, it is clearly seen that the twinning mode has a clear distinct influence on the forming limit diagram. As expected, the mechanical behavior of the material and its evolution have a strong influence on the predicted FLDs. Due to the twinning mechanism, the FLD predicted by the Barlat–Cazacu model is lower than that predicted by the more classical Barlat–von Mises model.

5 Conclusions

A reduced-order crystal plasticity method has been used in this work to model the plastic anisotropy while taking into account the twinning and slip mechanisms. The resulting two-surface constitutive model has been coupled with the M–K imperfection approach to predict the ductility limit of the AZ31B magnesium alloy. The proposed numerical tool has been developed as a generic Mathematica script so that it can be easily applied to other HCP materials. The predicted material behavior has been compared with experimental results, which allowed us to validate the proposed reduced-order model. Also, it has been shown that the combined yield surface (Cazacu–Barlat) has a great influence on the predicted FLD.

References

1. Aretz H (2004) Numerical restrictions of the modified maximum force criterion for prediction of forming limits in sheet metal forming. Modell Simul Mater Sci Eng 12(4):677–692
2. Barlat F, Lege DJ, Brem JC (1991) A six-component yield function for anisotropic materials. Int J Plast 7(7):693–712
3. Cazacu O, Plunkett B, Barlat F (2006) Orthotropic yield criterion for hexagonal closed packed metals. Int J Plast 22(7):1171–1194
4. Dudzinski D, Molinari A (1991) Perturbation analysis of thermoviscoplastic instabilities in biaxial loading. Int J Solids Struct 27(5):601–628
5. Eyckens P, Van Bael A, Van Houtte P (2011) An extended Marciniak-Kuczynski model for anisotropic sheet subjected to monotonic strain paths with through-thickness shear. Int J Plast 27(10):1577–1597
6. Hill R (1952) On discontinuous plastic states, with special reference to localized necking in thin sheets. J Mech Phys Solids 1(1):19–30
7. Hutchinson JW, Neale KW, Needleman A (1978) Sheet necking—I. Validity of plane stress assumptions of the long-wavelength approximation. In: Mechanics of sheet metal forming. Springer, Boston, pp. 111–126
8. Kondori B, Madi Y, Besson J, Benzerga AA (2018) Evolution of the 3D plastic anisotropy of HCP metals: experiments and modeling. Int J Plast (in press)

9. Madi Y, Benzerga A, Besson J (2017) Modeling the 3D plastic anisotropy of magnesium AZ31B alloy. In: Contributions to the foundations of multidisciplinary research in mechanics, proceedings of the XXIV international congress of theoretical and applied mechanics (ICTAM). IUTAM, pp 2730–2731
10. Marciniak Z, Kuczyński K (1967) Limit strains in the processes of stretch-forming sheet metal. Int J Mech Sci 9(9):609IN1613–612IN2620
11. Plunkett B, Lebensohn RA, Cazacu O, Barlat F (2006) Anisotropic yield function of hexagonal materials taking into account texture development and anisotropic hardening. Acta Mater 54(16):4159–4169
12. Rudnicki JW, Rice JR (1975) Conditions for the localization of deformation in pressure-sensitive dilatant materials. J Mech Phys Solids 23(6):371–394
13. Steglich D, Tian X, Besson J (2016) Mechanism-based modelling of plastic deformation in magnesium alloys. Eur J Mech-A/Solids 55:289–303
14. Swift H (1952) Plastic instability under plane stress. J Mech Phys Solids 1(1):1–18
15. Tong V, Wielewski E, Britton B (2018) Characterisation of slip and twinning in high rate deformed zirconium with electron backscatter diffraction. arXiv preprint arXiv:1803.00236
16. Wu SH, Song NN, Pires FMA, Santos AD (2015) Prediction of forming limit diagrams for materials with HCP structure. Acta Metall Sinica (English Letters) 28(12):1442–1451
17. Yu K, Li WX, Wang RC (2005) Plastic deformation mechanism of magnesium alloys. Chin J Nonferrous Metals 15(7):1081

Effect of Capacity Tightness on Performance of MOPSO Algorithm: Case of Multi-item Capacitated Lot-Sizing Problem

Hanen Ben Ammar(✉), Omar Ayadi, and Faouzi Masmoudi

Mechanics, Modeling and Manufacturing Research Laboratory, Sfax University, National Engineering School of Sfax (ENIS), Route de Sokra B.P, 1173-3038 Sfax, Tunisia
Benammar.hanen@hotmail.fr, omar.ayadi@yahoo.fr, faouzi.masmoudi@enis.tn

Abstract. This paper investigates the effect of capacity tightness on performance of multi-objective particle swarm optimization (MOPSO) algorithm in solving the multi-item capacitated lot-sizing problem with consideration of setup times and backlogging (MICLSP-SB). The considered problem is formulated as a multi-objective optimization model. The formulated model aims at simultaneously minimizing two objective functions. The first one seeks to minimize the total cost, which the sum of production, setup and backlogging costs. The second one seeks to minimize the total inventory level. Sensitivity analysis is performed on a set of generated problem instances. The capacity tightness factor is defined as the ratio between the required capacity and the total available capacity. Three levels of capacity tightness factor are considered for each problem instances. The metrics, which are used for evaluating the performances of the MOPSO algorithm, are number of Pareto solutions, spacing and computational time. Results of sensitivity analysis show a considerable impact of capacity tightness on performances of MOPSO algorithm in terms of number of Pareto solutions and computational time. This investigation offers to the decision makers a clear insight into the performances of MOPSO algorithm in solving the considered MICLSP-SB depending on problem features especially, the capacity tightness factor.

Keywords: Lot-sizing · Capacity tightness · Production planning · Multi-objective particle swarm optimization

1 Introduction

The multi-item capacitated lot-sizing problem labeled CLSP is a production-planning problem. The CLSP consists in finding production plans, which minimize production, inventory and setup costs while satisfying the demands for a set of items over a given planning horizon [3]. Since the CLSP with setup times belongs to NP-hard problems [1], a lot of approaches are used to solve this problem in the literature. Many of these studies used metaheuristic approaches such as harmony search [7], vibration damping

F. Chaari et al. (eds.), *Advances in Materials, Mechanics and Manufacturing*, Lecture Notes in Mechanical Engineering,
https://doi.org/10.1007/978-3-030-24247-3_22

optimization algorithm [8], non-dominated sorting genetic algorithm [11], multi-objective harmony search algorithm [3], non-dominated ranking genetic algorithm [6].

Particle swarm optimization (PSO) which is a population based-stochastic optimization technique is developed [4]. It has been successfully used for solving the lot-sizing problems in the literature. In this context, [10] utilized PSO for solving a stochastic dynamic lot-sizing problem with consideration of backlogging. They proved that the PSO algorithm could very efficient even with large-sized instances with respect to solution accuracy and computational time. A particle swarm optimization-based algorithm is developed by Pan et al. [9] for solving the multi-item capacitated production-planning problem. The authors demonstrated that the proposed algorithm outperforms commercial optimization software with respect to solution quality and execution time.

In recent years, [12] proposed an iterative improvement binary PSO (IIBPSO) approach for solving a very large capacitated multi-item multi-level lot-sizing problem. The proposed approach consists at finding a reasonable solution using binary PSO. Then, they applied an improved local search procedure in order to improve the obtained solution with binary PSO. The proposed approach IIBPSO outperforms the BPSO, the binary genetic algorithm (BGA) and iterative improvement BGA (IIBGA) in terms of solution quality.

Although many works in the literature focused on the application of PSO in solving lot-sizing problems, no one has investigated the impact of the problem in-stance features (e.g. capacity tightness) on the performance of PSO algorithm.

The purpose of this paper is to investigate the effect of capacity tightness on performances of multi-objective particle swarm optimization (MOPSO) in solving the multi-item capacitated lot-sizing problem with consideration of setup times and backlogging (MICLSP-SB).

The paper is organized as follows. Section 2 presents the problem formulation. Section 3 describes the multi-objective particle swarm optimization. In Sect. 4, the experimental results are detailed. Finally, Sect. 5 draws conclusions of this paper.

2 Problem Formulation

The presented model is a bi-objective optimization model for MCLSP-SB, which provides an optimal production plan over a given planning horizon while meeting demands with respect to the capacity constraints. The formulated model attempts to minimize simultaneously the total cost and the total inventory level.

The formulated model must satisfy the following assumptions:

- Demand of each item type is assumed to be deterministic
- Backlogging and inventory quantities are determined at the end of each period
- Backlogging and inventory quantities are null at the beginning and the end of the planning horizon
- Shortage is considered as backlogging.

Consider the following notations and parameters:

Indexes

i product index, i = 1, …, N,
t period index, t = 1, …, T

Parameters

D_{it}: Demand of item i at period t
bc_i: Unit backlogging cost of item i
c_i: Unit production cost of item i
s_i: Unit setup cost of item i
v_i: Required time to produce a unit of item i
st_i: Setup time of item i
M_{it}: Maximum production level of item i in period t
C_t: Available capacity in period t (in time units)

Decision Variables

X_{it}: Production quantity of item i in period t
I_{it}: Inventory quantity of item i at the end of period t
B_{it}: Backlogging quantity of item i at the end of period t
Z_{it}: Binary setup variable. Z_{it} = 1 if the resource is setup for item i in period t, and 0 otherwise.
I: Total Inventory Level
TC: Total Cost.

The developed mathematical model is presented below:

$$Min\ \ TC = \sum_{t=1}^{T}\sum_{i=1}^{N}[(c_iX_{it}) + (s_iZ_{it}) + (bc_iB_{it})] \tag{1}$$

$$Min\ I = \sum_{t=1}^{T}\sum_{i=1}^{N} I_{it} \tag{2}$$

Subject to

$$X_{it} + I_{i,t-1} + B_{it} = D_{it} + I_{it} + B_{i,t-1} \quad \forall t = 1,\ldots,T; \quad i = 1,\ldots,N \tag{3}$$

$$\sum_{i=1}^{N}(v_iX_{it} + st_iZ_{it}) \le C_t \quad \forall t = 1,\ldots,T \tag{4}$$

$$X_{it} \leq M_{it} Z_{it} \forall t = 1, \ldots, T; \quad i = 1, \ldots, N \tag{5}$$

$$I_{it} \times B_{it} = 0 \quad \forall i = 1, \ldots, N; \quad t = 1, \ldots, T \tag{6}$$

$$M_{it} = \min\left\{ \sum_{t'=t}^{T} D_{it'}, \frac{(C_{t'} - st_i)}{v_i} \right\} \quad \forall i = 1, \ldots, N \tag{7}$$

$$X_{it}, I_{it}, B_{it} \geq 0 \quad \forall t = 1, \ldots, T; \quad \forall i = 1, \ldots, N \tag{8}$$

$$Z_{it} \in \{0, 1\} \quad \forall t = 1, \ldots, T; \quad i = 1, \ldots, N \tag{9}$$

The first objective function (1) aims to minimize the total cost that is sum of production, setup and backlogging costs. The second objective function (2) tries to minimize the total inventory level. Constraints (3) details the flow balance equation between production, inventories, backlogging and demand for each item type at each period. Constraints (4) define the production capacity in each period. Constraints (5) present the production and setup linkage. Constraints (6) define the interdependence between inventory and backlogging levels for each item at each period. The upper bound for the lot size of each item type at each period is represented by constraints (7). Constraints (8) and (9) specify the decisions variables domains.

3 Multi-objective Particle Swarm Optimization

Particle swarm optimization (PSO) is a population-based stochastic optimization technique. The recent literature review stated that PSO becomes potential competitor of genetic algorithm that is mainly used to solve a wide variety of single as well as multi-objective optimization problems [2, 5]. Moreover, PSO shows a good performances in generating near-optimal solutions for lot-sizing problems [12].

In this paper, a multi-objective particle swarm optimization (MOPSO) is adopted in order to find the Pareto optimal solutions corresponding to the presented bi-objective model.

The main steps of the adopted MOPSO algorithm are as follows:

Step 1: Initialize the position P_i^0 of each particle of the population POP:
Step 2: Initialize the velocity of each particle on the POP $V_i^0 = 0$
Step 3: Evaluate each particle of the POP
Step 4: Save the non-dominated particles in an archive called repository (REP)
Step 5: Create the hypercubes of the search space, and using these hypercubes as a coordinates system for locating the particles
Step 6: Initialize the personal best position of each particle: $Pbest_i^t = P_i^t$
Step 7: Initialize counter loop t = 1
Step 8: Select REP^t from repository as global best position
Step 9: Compute the velocity of each particle i at iteration t + 1 by the means of the following expression:

$$V_i^{t+1} = W \times V_i^t + C_1 \times R_1 \times \left(Pbest_i^t - P_i^t\right) + R_2 \times C_2 \times \left(REP^t - P_i^t\right) \quad (10)$$

where W is the inertia weight, C_1 and C_2 are cognitive and social factors respectively. The parameters R_1 and R_2 are two random numbers in the range [0, 1].
Step 10: Compute the new position of each particle using the Eq. (11).

$$P_i^{t+1} = P_i^t + V_i^{t+1} \quad (11)$$

Step 11: Evaluate each particle of the POP
Step 12: Update the contents of repository and the localization of the particles within the hypercubes
Step 13: Update the personal best position
Step 14: Increment the loop counter t
Step 15: If stopping criterion (maximum number of iterations (Maxit)) is not reached, go to step 8, else, report the contents of the repository as final Pareto front.

4 Experimental Results

In this section, firstly, the considered instances are described. Then, obtained results are presented and analyzed.

4.1 Instances Description

Four problem instances are studied. The main features are presented in Table 1, where N and T represent the number of items and periods, respectively. The purpose of this work is to investigate the effect of the capacity tightness on the performances of the MOPSO algorithm in solving the MICLSP-SB. For this purpose, for each problem instances, the capacity tightness (CT) is considered at three levels as shown in Table 2.

Table 1 Considered instances

Parameters	N	T
P1	4	6
P2	8	10
P3	12	6
P4	10	12

The parameters used for MOPSO algorithm are given in Table 3. The MOPSO algorithm is coded in MATLAB R2103b software. Computational tests are performed on an Intel Core i5 CPU 3.2 GHz PC with 8.00 GB of RAM.

Table 2 Levels of capacity tightness

Capacity tightness CT	Level		
	0.75	0.85	0.95

Table 3 Parameters of MOPSO algorithm

Parameters	Value
NPop	75
Maxit	1000
NRep	100
W	1
C1	2
C2	2

4.2 Results and Discussion

The performances of the MOPSO algorithm are quantified by the following metrics:

- *Number of Pareto solutions*: measures the number of non-dominated solutions obtained with an algorithm. Larger value of this metric is preferred than the smaller one.
- *Spacing*: measures the uniformity of the distribution of the non-dominated solutions along the discovered Pareto front [13]. The small value of spacing metric is desirable.
- *Computational time*: measures the execution time for running an algorithm to find near optimal solutions.

Figure 1 shows the non-dominated solutions obtained with MOPSO algorithm for each problem instances with the three levels of capacity tightness.

The computational results of MOPSO algorithm in terms of number of Pareto solutions for solving the problem instances at different levels of capacity tightness reported in Fig. 2. As shown in this figure and Fig. 2, the performance of MOPSO algorithm in terms of number of Pareto solutions decreases with the increase of the capacity tightness level of the problem instance. In fact, the average number of Pareto solutions obtained with MOPSO for solving problem instances with a capacity tightness equal to 0.85 is 12% lower than the average one obtained with a capacity tightness equal to 0.75. Whereas, the average number of Pareto solutions obtained with MOPSO algorithm for solving problem instances with a capacity tightness equal to 0.95 is 56% lower than the average one obtained with a capacity tightness equal to 0.85. It can concluded that MOPSO algorithm has better performance in terms of number of Pareto solutions in solving the considered problem instances with a capacity tightness equal to 0.75 and 0.85 than the one with a capacity tightness equal 0.95. However, even though the number of Pareto solution of MOPSO algorithm for solving problem instance with capacity tightness level equal to 0.95 is less, it remains important for all considered instances and varies between 10 and 30 solutions in the Pareto front.

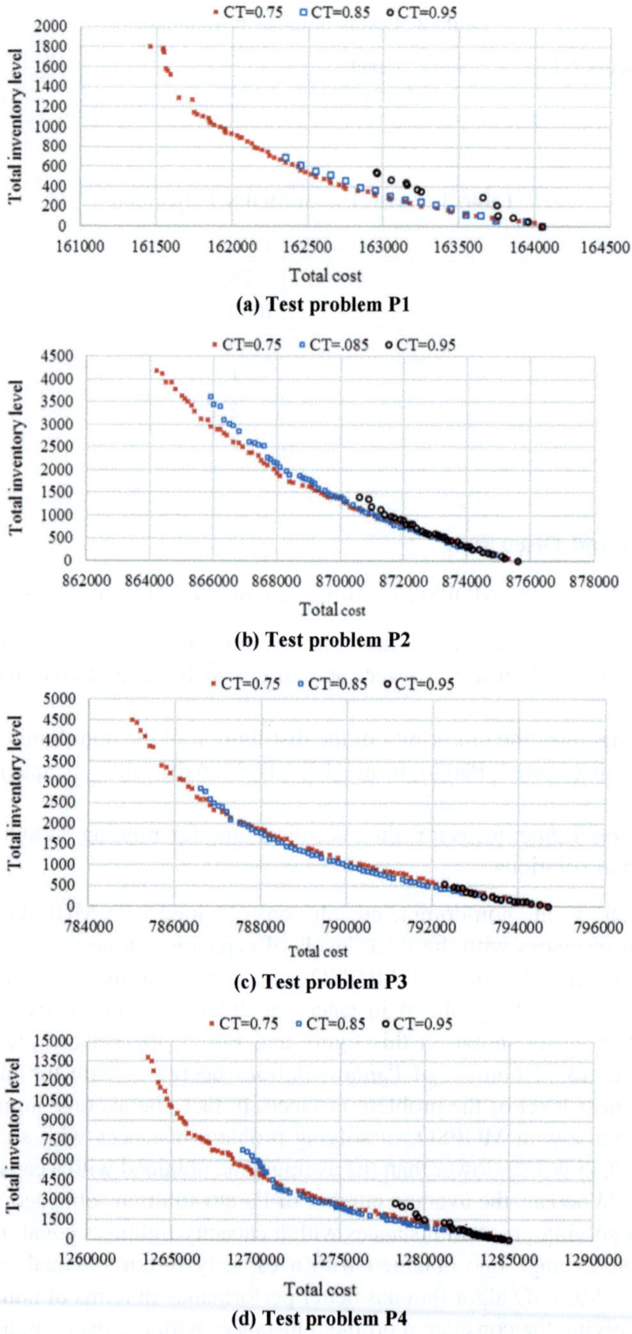

Fig. 1 Obtained Pareto front for considered instances in different levels of capacity tightness

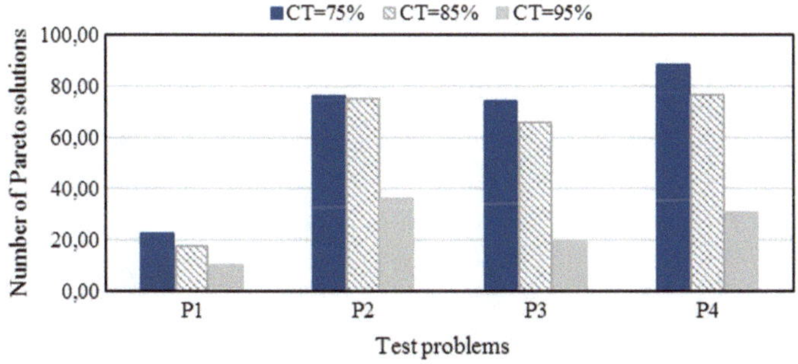

Fig. 2 Effect of capacity tightness on the performance of MOPSO in terms of number of Pareto solutions metric

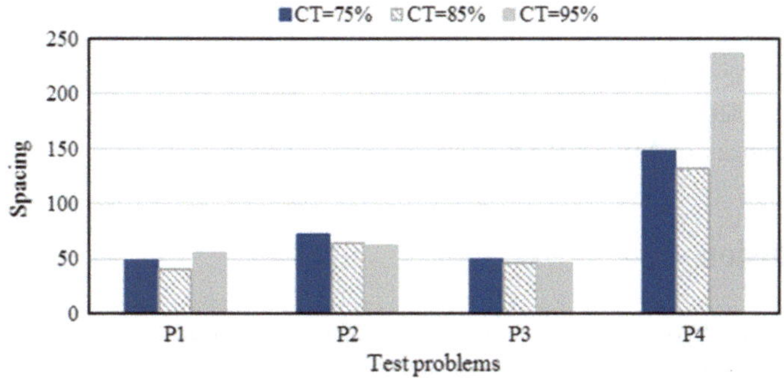

Fig. 3 Effect of capacity tightness on the performance of MOPSO in terms of spacing metric

Figure 3 shows the effect of the capacity tightness on performance of MOPSO algorithm in solving the different considered problem instances, regarding the spacing metric. It is clear from this figure and Fig. 1, that there is no significant effect on the performance of MOPSO algorithm for solving problem instances P1, P2 and P3 with different levels of capacity tightness. In addition, there is no significant effect of capacity tightness on performance of MOPSO algorithm in terms of spacing metric for solving problem instance P4 with the levels of capacity tightness equal to 0.75 and 0.85, respectively as shown in Fig. 3. Whereas, the spacing value obtained with MOPSO algorithm for problem instances P4 with capacity tightness equal to 0.95 is 78% greater than the one obtained with a capacity tightness equal to 0.85.

Figure 4 shows the effect of the capacity tightness on performance of MOPSO algorithm in terms of computational time for the different considered problem instances. According to this figure, the computational time of the MOPSO algorithm increases with increase of the capacity tightness level of the problem instance. In fact, the average computational time of MOPSO algorithm for solving problem instances with capacity tightness equal to 0.85 is 12% greater than the average one obtained with a capacity tightness equal to 0.75. Whereas, the average computational time of the

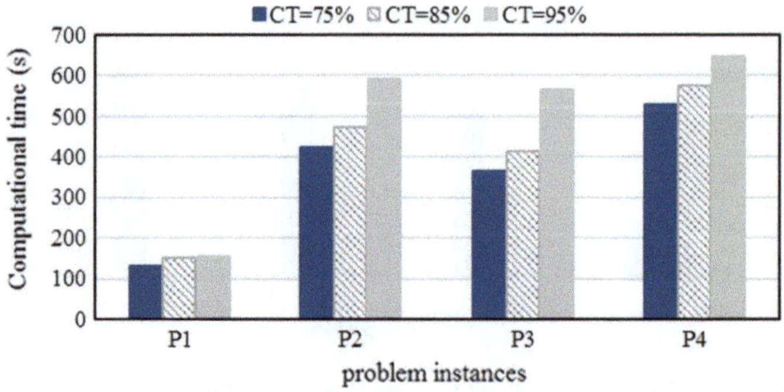

Fig. 4 Effect of capacity tightness on the performance of MOPSO in terms of CPUT metric

MOPSO algorithm for solving problem instances with capacity tightness equal to 0.95 is 24% greater than the average one obtained with a capacity tightness equal to 0.85.

Based on the sensitivity analysis, it can concluded that the capacity tightness has a significant impact on the performance of MOPSO algorithm in solving the considered problem instances of MICLSP-SB.

5 Conclusion

In this paper, an investigation of the impact of capacity tightness factors on the performances of MOPSO algorithm in solving a bi-objective planning model is proposed. The considered model attempts to minimize the total cost and total inventory level. A randomly generated problem instances, having three levels of capacity tightness, are considered. For each studied problem instance, while changing the capacity tightness level, the performances of the MOPSO algorithm in terms of the number of Pareto solutions and computational time varies significantly. Whereas, regarding the spacing metric, there is no significant impact on the performances of MOPSO algorithm. Based on the sensitivity analysis performed on the considered problem instances, it can be concluded that the capacity tightness feature has a significant impact on the performances of MOPSO algorithm in solving the considered MICLSP-SB.

In fact, the performance of MOPSO algorithm in terms of the number of Pareto solutions decreases with the increase of capacity tightness level. However, even though this performance decreases, it remains important and varies between 10 and 30 solutions in the final Pareto front for a problem instances with a capacity tightness equal to 0.95. Moreover, the computational time of MOPSO algorithm for solving the considered problem instances increases with the increase of capacity tightness level.

References

1. Chen W-H, Thizy J-M (1990) Analysis of relaxations for the multi-item capacitated lot-sizing problem. Ann Oper Res 26:29–72. https://doi.org/10.1007/BF02248584
2. Ding S, Chen C, Xin B, Pardalos PM (2018) A bi-objective load balancing model in a distributed simulation system using NSGA-II and MOPSO approaches. Appl Soft Comput J 63:249–267. https://doi.org/10.1016/j.asoc.2017.09.012
3. Hajipour V, Kheirkhah A, Tavana M, Absi N (2015) Novel Pareto-based meta-heuristics for solving multi-objective multi-item capacitated lot-sizing problems. Int J Adv Manuf Technol 80:31–45. https://doi.org/10.1007/s00170-015-6993-6
4. Kennedy J, Eberhart R (1995) Particle swarm optimization. In: IEEE international conference of neural networking, pp 1942–1948. https://doi.org/10.1109/ICNN.1995.488968
5. Lin JT, Chiu C-C (2018) A hybrid particle swarm optimization with local search for stochastic resource allocation problem. J Intell Manuf 29:481–495. https://doi.org/10.1007/s10845-015-1124-7
6. Mehdizadeh E, Hajipour V, Mohammadizadeh MR (2016) A bi-objective multi-item capacitated lot-sizing model: two Pareto-based meta-heuristic algorithms. Int J Manage Sci Eng Manag 11:279–293. https://doi.org/10.1080/17509653.2015.1086965
7. Mehdizadeh E, Kivi AF (2014) Three meta-heuristic algorithms for the single-item capacitated lot-sizing problem. Int J Eng 27:1223–1232. https://doi.org/10.5829/idosi.ije.2014.27.08b.08
8. Mehdizadeh E, Kivib AF (2014) Three metaheuristic algorithms for solving the multi-item capacitated lot-sizing problem with product returns and remanufacturing. J Optim Ind Eng 16:41–53
9. Pan T, Zhang Z, Cao H (2014) Collaborative production planning with production time windows and order splitting in make-to-order manufacturing. Comput Ind Eng 67:1–9. https://doi.org/10.1016/j.cie.2013.10.006
10. Piperagkas GS, Konstantaras I, Skouri K, Parsopoulos KE (2012) Solving the stochastic dynamic lot-sizing problem through nature-inspired heuristics. Comput Oper Res 39:1555–1565. https://doi.org/10.1016/j.cor.2011.09.004
11. Rezaei J, Davoodi M (2011) Multi-objective models for lot-sizing with supplier selection. Int J Prod Econ 130:77–86. https://doi.org/10.1016/j.ijpe.2010.11.017
12. Sahithi VV, Krishna PS, Lalithkumar K, Rao CSP (2015) An iterative improvement search and binary particle swarm optimization for large capacitated multi item multi level lot sizing (CMIMLLS) problem. Int J Eng Res Gen Sci 3:1092–1106
13. Suo X, Yu X, Li H (2017) Subset simulation for multi-objective optimization. Appl Math Model. https://doi.org/10.1016/j.apm.2017.02.005

Rayleigh Damping Coefficients Identification Using the Wavelet Transform on Two Stage Gear System

Nourhaine Yousfi[1,2](✉), Bacem Zghal[1], Ali Akrout[1,2], Lassaad Walha[1], and Mohamed Haddar[1]

[1] Mechanics, Modelling and Production Research Laboratory (LA2MP), Engineering National School of Sfax, University of Sfax, Sfax, Tunisia
nourhainegem@gmail.com, bacem.zghal@isgis.rnu.tn, {ali_akrout2005,walhalassaad}@yahoo.fr, mohamed.haddar@enis.rnu.tn

[2] Engineering National School of Tunis, University of Tunis El-Manar, Tunis, Tunisia

Abstract. Damping plays an important role in the simulation of the mechanical system. Various damping models are used such us non-viscous damping, coulomb damping, and hysteresis damping. For mathematical convenience, Rayleigh damping is usually used in a gear system to model damping. Rayleigh damping coefficients (RDCs) can be identified using the frequencies when the degree of freedom of the system is low. However, for complex systems, problems of selecting the frequencies of the RDCs are presented. The classical methods used only a constant damping ratio for all modes can underestimate the dynamic response. Continuous wavelet transforms (CWT) method is recently used to identify the modal parameters such as the natural frequencies and the damping ratios. This study presents initial research into the use of the CWT method to select the optimal frequencies of the Rayleigh damping formulation to identify RDCs. The RDCs are identified using the modes contributing to the dynamic response of the two-stage gear system. The modes that remarkably affect the dynamic responses are determined on the basis of the modes corresponding to the maximum values of the wavelet spectrum. The proposed method is validated using the simulated responses of a two-stage gear system and compared with the classical methods.

Keywords: Rayleigh damping formulation · Optimum frequencies · Continuous wavelet transforms · Dominant frequency

1 Introduction

Non-viscous damping model is usually used on gear system analysis. Recently, we presented a new technique to identify non-viscous damping on one stage system using the integral method [1]. However, Rayleigh damping model is the most used formulation of damping. In the structural dynamics, many procedures are used to select RDCs such us the method presented by Chowdhury and Dasgupta [2] for systems with

F. Chaari et al. (eds.), *Advances in Materials, Mechanics and Manufacturing*, Lecture Notes in Mechanical Engineering,
https://doi.org/10.1007/978-3-030-24247-3_23

large degrees of freedom but their study neglects the influence of input motion to select the optimal frequencies of RDCs. Zhe et al. [3] developed a new method to calculate Rayleigh damping coefficients based on the spectrum of the dynamic response. Kwok et al. [4] illustrated that the use of 1st and five times that frequency can be used to estimate RDCs. The target frequencies are established through a process by which linear time domain and frequency domain solutions are matched.

For the gear system analysis, a constant damping ratio is always used to identify RDCs [5]. However, this type of the approach has some limitations and can underestimate the dynamic response. Therefore, the damping ratios should be identified and the optimal frequencies of the Rayleigh damping formulation must be selected using an improved procedure. So, the CWT method is used to select optimum modes of the Rayleigh damping formulation that contribute to the dynamic responses from the dominant modes of the wavelet plot where the high energy is concentrated.

This study is organized as follows. Section 2 introduces the classical Rayleigh damping formulation. Section 3 presents the proposed calculation method of Rayleigh damping coefficients. Section 4 provides the validation of the proposed method using a numerical example of two-stage gear system.

2 Current Rayleigh Damping Formulation

On gear systems, the Rayleigh damping matrix [C] is derived from a combination of the mass matrix and the stiffness matrix [5]:

$$[C] = \alpha[M] + \beta[K] \tag{1}$$

where [M] = mass matrix, [K] = stiffness matrix. The coefficients α and β in Eq. (2) are defined using a constant damping ratio ξ for all modes, the first frequency f_0 and the second frequency f_1 as follows [4]:

$$\alpha = \xi \frac{4\pi(f_0 f_1)}{f_0 + f_1}; \quad \beta = \xi \frac{1}{\pi(f_0 + f_1)} \tag{2}$$

This formulation can underestimate the dynamic response and presents some limitations.

3 Proposed Calculation Method of Rayleigh Damping Coefficients

In the original formulation proposed by Rayleigh and Lindsay [4], α and β coefficients in Eq. (2) can be determined using two modes m and n:

$$\begin{cases} \alpha = \frac{2W_k W_n}{W_n^2 - W_k^2}(\xi_k w_n - \xi_n w_k) \\ \beta = \frac{2(\xi_n W_n - \xi_k W_k)}{W_n^2 - W_k^2} \end{cases} \tag{3}$$

where w_n and w_k correspond to the frequency of the n-th mode and k-th mode respectively, ξ_n and ξ_k are the damping ratios. So, the modes w_n and w_k should be selected to find the RDCs. This study uses this formulation on gear system to model the damping matrix by specifying the desired damping ratios of two different modes using the wavelet transform and to present the limits of the classical Rayleigh damping formulation which considers a constant damping ratio.

The continuous wavelet transforms (CWT) based on damping identification methods is used to identify damping ratios on two stage gear system in this first step then, to select the frequencies of RDCs based on the dominant frequencies. The CWT-based damping-identification procedure is composed by the following steps. Initially the signal of the gear system should be transformed at time-frequency plot which presents the variation of this amplitude with the time and the scale s using the following equation [6]:

$$\begin{cases} X_{\psi}(a,b) = \int_{-\infty}^{+\infty} x(t)\psi^{*}_{a,b}(t)dt \\ \psi_{a,b}(t) = \frac{1}{\sqrt{a}}\psi\left(\frac{t-b}{a}\right) \end{cases} \quad (4)$$

where $\Psi^{*}(t)$ are the complex conjugate of $\Psi(t)$. Using the modulus of the wavelet coefficients $|W_x(a,b)|$.

The second step is the envelope extraction of each mode separately. The third step is to identify the damping ratio using the slope of the straight line of the semi logarithmic plot of wavelet modulus [6]:

$$\ln|w_x(a_0,b)| = \xi\omega_n b + \ln\left(\frac{\sqrt{a_0}}{2}B|\psi^{*}|(a_0\omega_d)\right) \quad (5)$$

The last step of the CWT-based damping-identification is to determine the dominant frequencies, from which the optimal frequencies of the Rayleigh damping formulation are selected. The proposed procedure uses the 1st mode for w_n and defines w_k based on the range of frequency at which the high amplitude of wavelet transform is concentrated.

4 Numerical Example and Validation of the Proposed Algorithm

4.1 Numerical Example

The dynamic model of a two-stage gear system is studied with the parameters presented at Table 1 [5]. Gears are modelled by concentrated masses, Bearings are modelled by linear springs (Fig. 1). The dynamic model studied is characterized by 12 degrees of freedom and can be defined by

Table 1 Parameters of the studied two-stage gear system

Material: 42CrMo4	$\rho = 7860\,kg/m^3$
Motor torque (N m)	Cm = 10
Bearings stiffnesses (N/m)	$kx_i = ky_i = 10^6$
Torsional shaft flexibilities (N m/rad)	$k\theta_i = 10^5$
Teeth module (m)	$m = 4 \times 10^{-3}$
Teeth number	Z1 = 30; Z2 = 45, Z3 = 30; Z4 = 45
Average mesh stiffness (N/m)	$k_1moy = k_2moy = 3 \times 10^8$

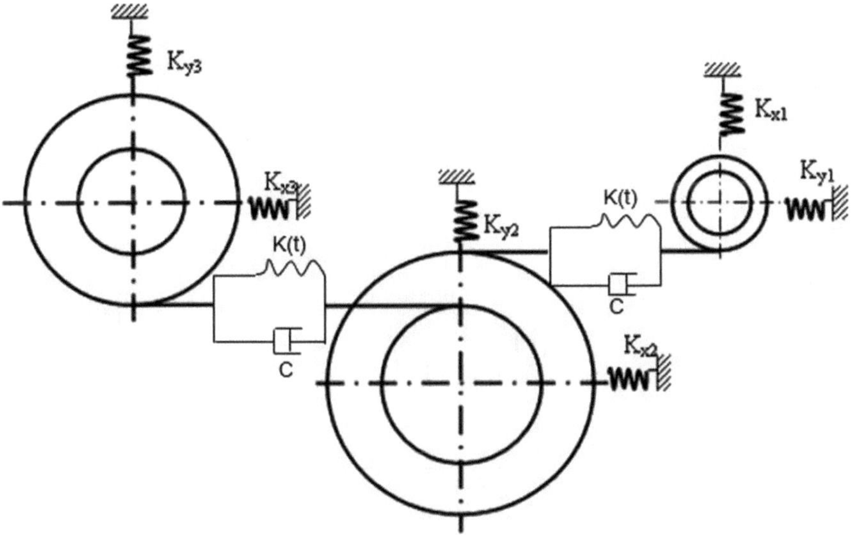

Fig. 1 Dynamic model of the two-stage spur gear system [5]

$$\{q\} = [x_1, y_1, x_2, y_2, x_3, y_3, \theta_m, \theta_1, \theta_2, \theta_3, \theta_4, \theta_r] \tag{6}$$

where xj and yj are the bearings displacements, (i = 1, …, 4) are the dynamic angular displacements the of the gears, θ_m and θ_r are the dynamic angular displacements of motor and brake.

The lagrange formulation was used to formulate the differential equation governing the system motion of the two stage gear:

$$[M]\{\ddot{q}\} + [C]\{\dot{q}\} + ([Ks] + [K(t)])\{q\} = \{F_0\} \tag{7}$$

[M] is the mass matrix. [Kmoy] is the average stiffness matrix of the system is of a size 12 × 12 [5]. Rayleigh damping [C] is of a size 12 × 12 given by

$$[C] = \alpha[M] + \beta[K_{moy}] \tag{8}$$

The two teeth deflections $\delta_1(t)$ and $\delta_2(t)$ of each stage gear are defined by

$$\delta_1(t) = (x_1 - x_2)\sin\alpha_1 + (y_1 - y_2)\cos\alpha_1 + r_1\theta_1 - r_2\theta_2$$
$$\delta_2(t) = (x_2 - x_3)\sin\alpha_2 + (y_2 - y_3)\cos\alpha_2 + r_3\theta_3 - r_4\theta_4 \tag{9}$$

The simulated responses correspond to the input first teeth deflection and the input second teeth deflection are obtained using step by step iterative Newmark method and presented in Fig. 2. Our objective is to identify the RDCs from only the simulated responses.

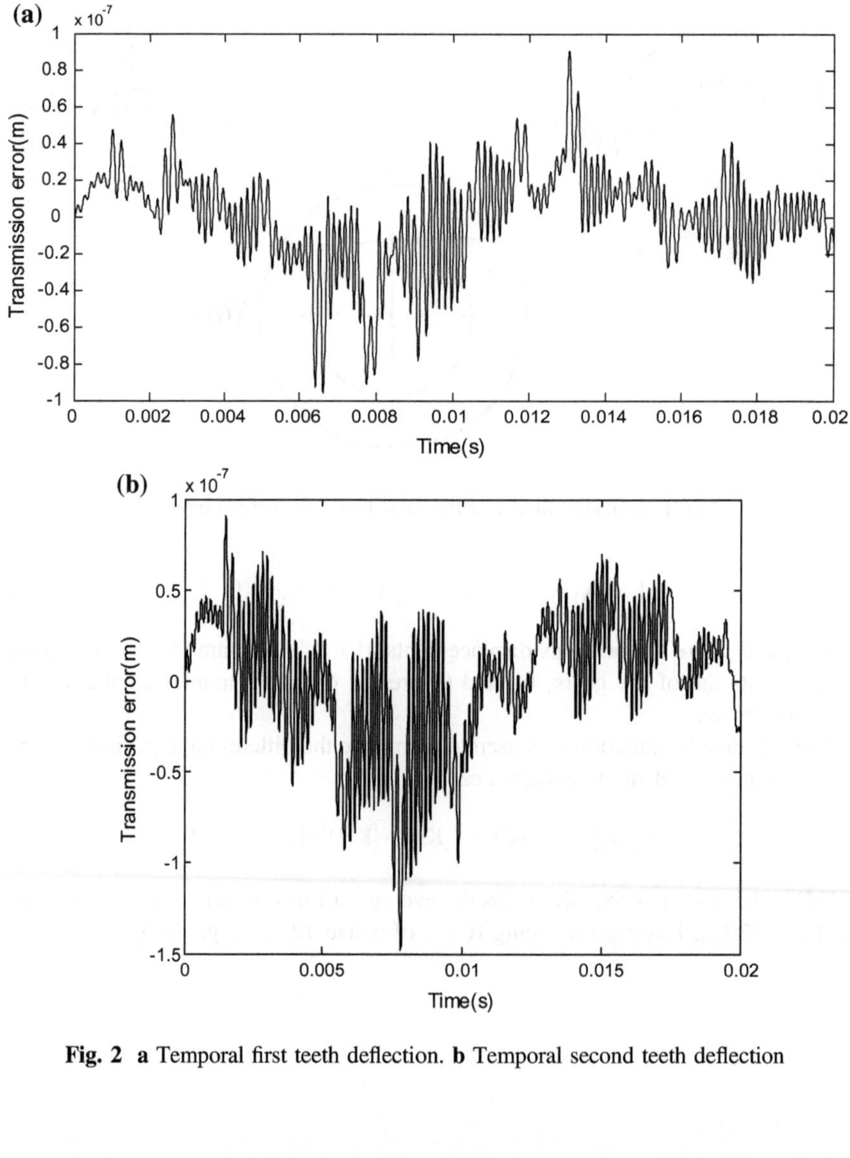

Fig. 2 **a** Temporal first teeth deflection. **b** Temporal second teeth deflection

4.2 Validation of the Proposed Method

4.2.1 Identification of the Damping Ratios

The procedure presented at Sect. 3 is used to identify the damping ratios of each stage separately. So, Fig. 3 shows the energy spectrum distribution in the time-scale (frequency) domain of the first stage and the second stage. For the first stage, seven values of the dilatation parameter are predominant in this plot. The 5th mode corresponds to the dominant frequency where high energy is concentrated

The wavelet transform is used to identify only the frequencies that are localized on the wavelet ridge where high energy is concentrated (high energy frequencies) on the basis of the modes corresponding to the maximum values of the wavelet spectrum. The wavelet envelope is presented in fig. 4.The wavelet envelope of each mode can be extracted from the wavelet plot by a slice parallel to the time axis through each frequency w_i for i = 1, …, 7 in the frequency axis which can be used to estimate the corresponding damping ratio. The identified modal parameters for the seven first modes are presented at Table 2. For the second stage, seven values of the dilatation parameter are predominant. The 4th mode corresponds to the dominant frequency and the same steps are used to found the modal parameters of the high energy frequencies presented at Table 3.

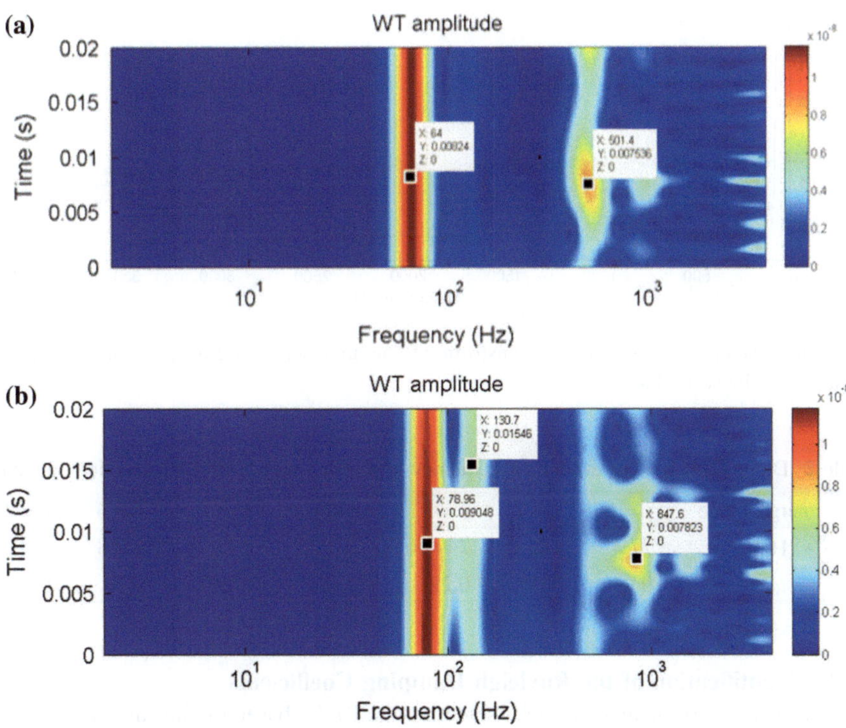

Fig. 3 **a** Wavelet plot of the first stage. **b** Wavelet plot of the second stage

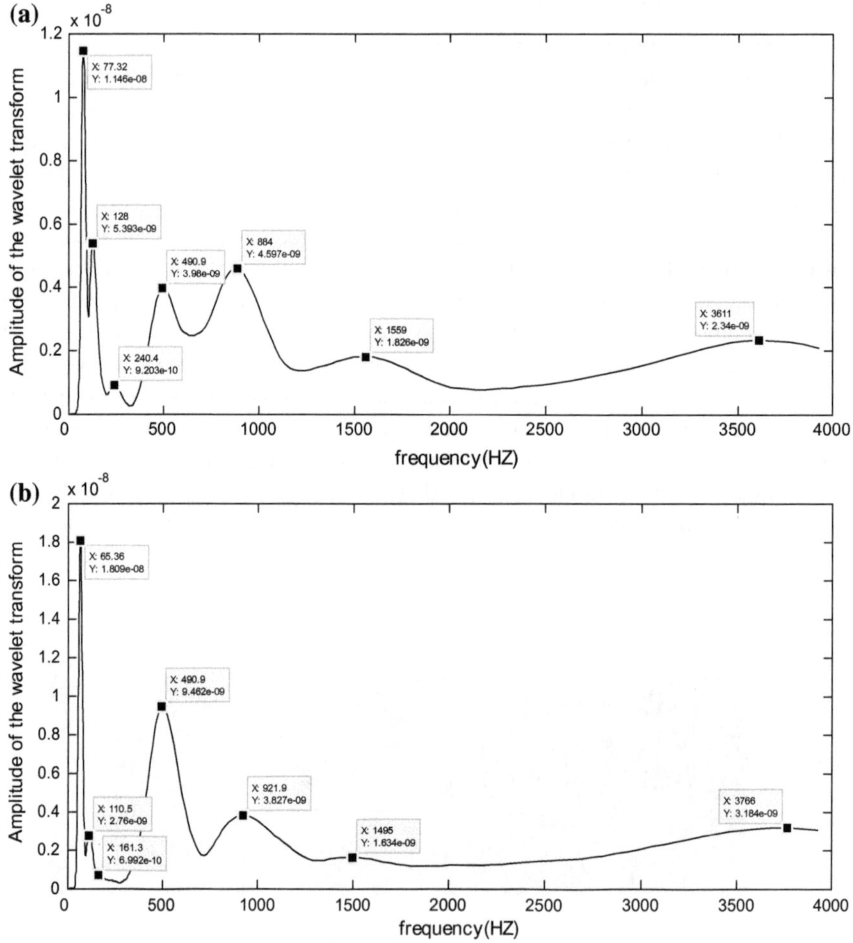

Fig. 4 **a** Envelope of the wavelet transform of the first stage. **b** Envelope of the wavelet transform of the second stage

Table 2 Damping ratios correspond to frequency of the first stage using wavelet demodulation

High energy mode	i = 1	i = 2	i = 3	i = 4	i = 5	i = 6	i = 7
$f_i = \frac{W_i}{2\pi}$ (Hz)	77.3	128	240.36	501.35	883.98	1558	3610
ς_i	0.04	0.01	0.22	0.186	0.009	0.1	0.14

4.2.2 Identification of the Rayleigh Damping Coefficients

This paper develops an algorithm for selecting the RDCs based on the calculated modal parameters obtained in the previous section using the wavelet transform. The selected coefficients using CWT method are closely compared to the results obtained using the classical Rayleigh formulations. The methods widely used and the proposed calculation

Table 3 Damping ratios correspond to frequency of the second stage using wavelet demodulation

High energy mode	i = 1	i = 2	i = 3	i = 4	i = 5	i = 6	i = 7
$f_i = \frac{w_i}{2\pi}$ (Hz)	64	115	168.18	501	921.9	1494.507	3678.64
ς_i	0.0008	0.01	0.08	0.0007	0.05	0.021	0.002

method of Rayleigh damping coefficients can be summarized at Table 4. Thus, the use of the proposed Rayleigh damping with the selected frequencies provides an exact match to the RDCs. The interpretation of Table 4 has been made. The coefficients α and β selected and obtained using the T.O method presented in Table 4 (Case 5) are compared to the results obtained using classical Rayleigh formulations (case 1, case 2, case 3, case 4 presented at Table 4).

Figures 5 and 6 compare the input teeth deflection of the two stages of the gear system with the obtained response using different optimal frequencies. As it shown, the conventional Rayleigh damping formulation underestimate the response and a good estimation is obtained when using w_k as the first mode and w_n as the dominant

Table 4 Rayleigh damping coefficients obtained by different methods for the input first teeth deflection

Methods	α	β	Involved orders
Case 1	0	0.001109	1
Case 2	4.4946	0.001861	1, 3
Case 3	2.7039	0.0005741	1, 4
Case 4	10.36	8.3965e−6	1, 5
Case 5	9.94	7.91e−5	1, 7

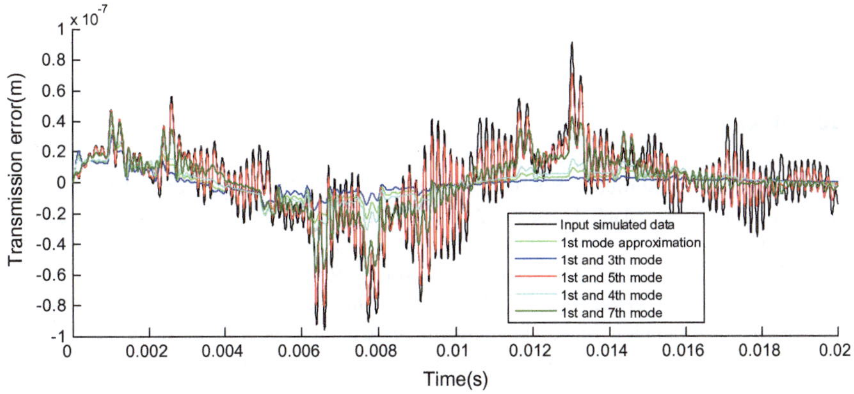

Fig. 5 First teeth deflection uses first natural mode approximation as well as proposed full Rayleigh viscous damping formulation

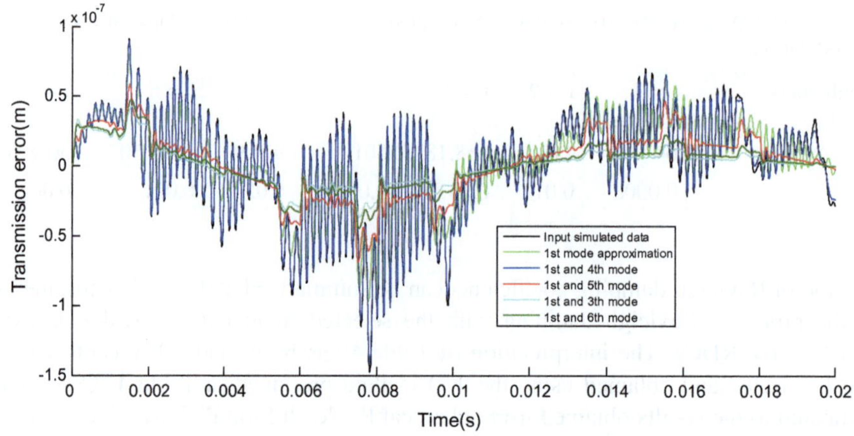

Fig. 6 Second teeth deflection uses first natural mode approximation as well as proposed full Rayleigh viscous damping formulation

frequency (higher mode) which corresponds the maximum value of the amplitude of the wavelet transform.

5 Conclusion

In the traditional method, the damping ratio is considered as constant for all modes to identify the RDCs in the gear system. As a result, the dynamic response can be underestimated. In the proposed method, the dominant modes that influence the dynamic responses are identified in the basis of the continuous wavelet transform method. Then, the optimum frequencies are selected from the dominant modes to obtain the RDCs. Comparisons with the simulated time domain response demonstrate that the proposed procedure is very efficient in selecting the optimum frequencies of the Rayleigh damping formulation from the first teeth deflection and the second teeth deflection of the gear system.

References

1. Yousfi N, Zghal B, Akrout A, Walha L, Haddar M (2018) Damping models identification of a spur gear pair. Mech Mach Theory 122:371–388. https://doi.org/10.1016/j.mechmachtheory.2018.01.002
2. Chowdhury I, Dasgupta S (2003 Jan) Computation of Rayleigh damping coefficients for large systems. Electron J Geotech Eng 138:324–336. https://www.researchgate.net/publication/284308903_Computation_of_Rayleigh_Damping_Coefficients_for_Large_Systems
3. Zhe L, Gongxian W, Yong H (2015) Application of improved calculation method of Rayleigh damping coefficients to seismic response analysis on quay crane structure. J South China Univ Technol Nat Sci Ed 43:103–109. https://doi.org/10.1155/2017/2046345

4. Kwok Annie OL, Stewart Jonathan P, Hashash Youssef MA, Matasovic N (2007) Use of exact solutions of wave propagation problems to guide implementation of nonlinear seismic ground response analysis procedures. J Geotech Geo-environ Eng 133:1385–1398.https://doi.org/10.1061/(ASCE)1090-0241(2007)133:11(1385
5. Walha L, Fakhfakh T, Haddar M (2009) Nonlinear dynamics of a two-stage gear system with mesh stiffness fluctuation, bearing flexibility and backlash. Mech Mach Theory 44:1058–1069. https://doi.org/10.1016/j.mechmachtheory.2008.05.008
6. Le T-P, Argoul A (2004) Continuous wavelet transform for modal identification using free decay response. J Sound Vib 277:73–100. https://doi.org/10.1016/j.jsv.2003.08.049

Optimizing Cutting Conditions in Single Pass Face Milling for Minimum Cutting Energy, Time, Cost, and Surface Roughness

Anoire Ben Jdidia[1(✉)], Taissir Hentati[1], Alain Bellacicco[2], Mohamed Taoufik Khabou[1], Alain Rivier[2], and Mohamed Haddar[1]

[1] Laboratory Mechanics, Modeling and Production, National Engineering School of Sfax (ENIS), BP 1173-3038, Sfax, Tunisia
benjdidia.anoire@gmail.com, taissirhentati@yahoo.fr, mtkhabou@hotmail.com, mohamed.haddar@enis.rnu
[2] QUARTZ Laboratory High, Institute of Mechanic of Paris(SUPMECA), 3 Rue Fernand Hainaut, 93400 Saint-Ouen Cedex, France
{alain.bellacicco,alain.riviere}@supmeca.fr

Abstract. Technology evolution and the demand of modern life have led to more using for machine tools which are the basic energy consumption devices in manufacturing. Subsequently, CO_2 emissions in the atmosphere will increase, causing several climate changes such as the greenhouse effect. As the resources and energy in the earth are limited and getting fewer and fewer, sustainable manufacturing is gaining more and more attention to produce the same product with less negative environmental impacts. In this paper, a mono-objective optimization for sustainable manufacturing is presented. Such approach needs a balance between economic and ecological aspects. Thus, the objective of this work is machining product with less environmental impacts by minimizing consumed energy with respect to technological and economic constraints. The consumed energy is modelled based on the dynamic behavior of the cutting forces. A case study of single pass of face milling operation is carried out using the particle swarm optimization tool. The surface quality is adopted as an objective in this work. Three decision variables are taken into account during the resolution such as rotational speed, axial depth of cut and feed per tooth. Results show that the proposed optimization model has a great efficiency to find a trade-off between the four objective functions in order to minimizing them.

Keywords: Sustainable manufacturing · Optimization · Particle swarm · Consumed energy

1 Introduction

In manufacturing, the machining process is the main electrical energy consumer [7]. In fact, the CNC machining has an important effect on environment due to the high level of electrical energy consumption [3] and global warming [10]. Thus, the reduction of the consumed energy by the machining process is important [6]. For this reason several works aim to study the relationship between the electrical energy demanded and cutting

F. Chaari et al. (eds.), *Advances in Materials, Mechanics and Manufacturing*, Lecture Notes in Mechanical Engineering,
https://doi.org/10.1007/978-3-030-24247-3_24

parameters during machining. For example, Luan et al. [9] utilized the response surface method (RSM) to study the effect of cutting parameters (cutting speed, axial depth of cut, radial depth of cut and feed per tooth) on the consumed energy during a face milling operation. The obtained results minimize the consumed energy and ameliorate the surface roughness. Wang et al. [12] used genetic algorithm to find optimum values of cutting parameters in case of high speed milling process in order to achieve maximum machining efficiency. Jang et al. [5] adopted particle swarm algorithm to obtain optimum cutting parameters that reduce energy consumption in milling operation case and minimal lubrication case. Li et al. [8] presented a resolution of a multi objective problem of energy efficiency and cutting time in case of milling process based on Tabu Search algorithm (TS). Results show that the radial and axial depths of cut are the significant parameters on the consumed energy while the spindle rotational speed is the most significant on the cutting time. Alberteli et al. [1] presented an optimization of both consumed energy and treatment time of a face milling operation. Firstly, mono variable optimization considering only the cutting speed is performed. Secondly multi variable optimization, using a multi-dimensional exhaustive enumeration method considering the axial depth of cut, feed per tooth and cutting speed, to minimize both the energy consumption and the production time, is performed. Tapoglou et al. [11] have elaborated a novel approach in order to ameliorate the energy efficiency of machine tools based on online cutting conditions optimization.

The common point between the backgrounds of developed works described above that they all strived to minimize the consumed energy by the milling machine tool. However, the time variation of the milling forces during the removing material process as well as the incorporation of the surface roughness as an objective during the optimization of the consumed energy is neglected. Thus, the aim of this work is to develop a new model of face milling machining energy optimization by considering cutting time, surface roughness and cutting cost factors.

2 Objective Functions

The objective function can be modeled as the sum of four objective functions describing cutting time; cutting consumes energy, machining cost and surface quality as described in the above sections.

2.1 Cutting Time

The required time to remove material is calculated using the following equation:

$$f_1 = t_{machining} = \frac{L + d_a}{\Omega f_z N} \tag{1}$$

where L is the workpiece length, Ω is the spindle rotational speed, f_z is the feed per tooth, N is the tool teeth number and d_a is the approach distance calculated as following:

$$d_a = D/2 - \sqrt{\left(\frac{D}{2}\right)^2 - \left(\frac{a_e}{2}\right)^2} \tag{2}$$

where a_e is the radial depth of cut and D is the tool diameter.

2.2 Machining Energy

The mathematical model of cutting energy is presented in the next equation:

$$E_{machining} = \int_0^{t_{machinning}} P_{machinning}(t)dt \tag{3}$$

where $P_{machining}(t)$ is the variable power consumed by the machining system (spindle and axis feed) at the tool tip to remove material which can be estimated by the next model shown in equation:

$$f_2 = E_{machining} = \int_0^{t_{machining}} P_{machinning}(t)dt = \int_0^{t_{machining}} \left(F_t(t)\ V_c + F_f(t)\ V_f\right) dt \tag{4}$$

where V_c and V_f are respectively the cutting speed and the feed rate, $F_t(t)$ and $F_f(t)$ are respectively the tangential and the feed components of the cutting force. These two forces are variable and their values change with time due the non-linearity of the milling operation. They are calculated in two steps: firstly we calculate the differential tangential $dF_{t,i}$ radial $dF_{r,i}$ and axial $dF_{r,i}$ components for the ith tooth which are expressed as a nonlinear function of varying chip load $h(\Phi_i)$ expressed as following [2]:

$$\begin{cases} dF_{t,i}(\Phi_i(t)) = g(\Phi_i(t))k_t a_p h(\Phi_i(t)) \\ dF_{r,i}(\Phi_i(t)) = k_r g(\Phi_i(t))k_t a_p h(\Phi_i(t)) \\ dF_{a,i}(\Phi_i(t)) = k_a g(\Phi_i(t))k_t a_p h(\Phi_i(t)) \end{cases} \tag{5}$$

where k_t, k_r and k_a are the specific pressure of the cutting force considered as constants, a_p and f_z are respectively the axial depth of cut and the feed per tooth and $g(\Phi_i(t))$ is a function describing whether the ith tooth is active or not. It is expressed as following:

$$g(\Phi_i(t)) = \begin{cases} 1, & \Phi_{st} \leq \Phi_i(t) \leq \Phi_{ex} \\ 0, & else \end{cases} \tag{6}$$

with Φ_{st} and Φ_{ext} are respectively the cutter entry and exit angles.

The variable chip generated during the machining phase is composed of two components: static h_s and dynamic h_d caused by the instantaneous angular position of the ith tooth $\Phi_i(t)$.

$$h(\Phi_i(t)) = \underbrace{f_z \sin(\Phi_i(t))}_{h_s} + \underbrace{(u_x(t) - u_x(t-\tau))\sin(\Phi_i(t)) + (u_y(t) - u_y(t-\tau))\cos(\Phi_i(t))}_{h_d} \tag{7}$$

where $\Phi_i(t)$ is modelled as following:

$$\Phi_i(t) = \Omega t + (i-1)\,\Phi_p \tag{8}$$

where Φ_p is the tooth spacing angle.

For a face milling process, the cutting forces components acting on the workpiece on feed direction X, on normal direction Y and on axial direction Z are obtained from the next equilibrium relation:

$$\begin{Bmatrix} dF_{x,i}(\phi_i(t)) \\ dF_{y,i}(\phi_i(t)) \\ dF_{z,i}(\phi_i(t)) \end{Bmatrix} = \begin{bmatrix} -\cos(\phi_i(t)) & -\sin(\phi_i(t)) & 0 \\ \sin(\phi_i(t)) & \cos(\phi_i(t)) & 0 \\ 0 & 0 & 1 \end{bmatrix} \begin{Bmatrix} dF_{t,i}(\phi_i(t)) \\ dF_{r,i}(\phi_i(t)) \\ dF_{a,i}(\phi_i(t)) \end{Bmatrix} \tag{9}$$

The total cutting force components in the X, Y and Z directions are computed by summing the elementary cutting force components exerted by all tooth. It can be expressed:

$$F_c(t) = \begin{Bmatrix} F_x(t) \\ F_y(t) \\ F_z(t) \end{Bmatrix} = \begin{Bmatrix} \sum_{i=1}^{N} dF_{x,i}(\phi_i(t)) \\ \sum_{i=1}^{N} dF_{y,i}(\phi_i(t)) \\ \sum_{i=1}^{N} dF_{z,i}(\phi_i(t)) \end{Bmatrix} \tag{10}$$

To estimate these forces components, a resolution of the differential equation of motion of a spindle flexible structure using the finite element method [4], as shown in Eq. (11), is elaborated.

$$[M_b]\{\ddot{q}\} + 2\Omega[G_b]\{\dot{q}\} + \left(\left[K_b - \Omega^2[C_b]\right]\right)\{q\} = \{F_c(t,q)\} \tag{11}$$

where $[M_b]$, $[G_b]$, $[K_b]$ and $[C_b]$ are respectively the mass, the gyroscopic, the stiffness and the centrifugal matrices. The vector $\{q\}$ denotes the degrees of freedom vector caused by elastic movements and associated to different nodes. The second member constitutes the total cutting force.

2.3 Surface Quality

The quality of the surface is described by the roughness which is adopted as a function to minimize in our work and it is modeled as following:

$$f_3 = kV_c^{x_1} f_z^{x2} a_p^{x3} \tag{12}$$

where x_1, x_2, x_3 and k are constants depending on workpiece and tool material.

2.4 Machining Cost

The machining cost is calculated as a sum of machine cost, tool cost and energy cost as expressed:

$$f_4 = C_{total} = k_0 t_{machining} + k_e E_{machning} + k_t \frac{t_{machining}}{T} \tag{13}$$

where k_0 is the machine cost during the cutting phase, k_e is the cutting energy cost, k_t is the tool cost and T is the tool life modelled by [13]:

$$T = \left(\frac{C_T D^{b_v}}{V_c f_z^{u_v} a_p^{e_v} a_e^{r_v} z^{n_v}} \right)^{1/x_v} \tag{14}$$

where b_v, u_v, e_v, r_v, n_v, C_T and x_v are constants.

During the optimization of the objective functions, some constraints must be satisfied. In the next section, we describe those constraints.

3 Constraints

3.1 Cutting Power

The cutting parameters values should verify the condition on the available power. In fact, the machining consumed power must be lower than the maximum power available on the spindle machine P_{max} as shown:

$$g_1 = \frac{k_s a_p f_z N V_c}{60{,}000 \Pi D} \le P_{max} \tag{15}$$

where k_s is a specific pressure of the cutting force.

3.2 Cutting Force

The cutting force applied by the cutter tool on the workpiece must be lower than the maximal one that can be supported by the cutter tool. So, a constraint on the cutting force should be taken into account as following:

$$g_2 = \frac{k_s a_p f_z N}{\Pi D} \leq F_{max} \tag{16}$$

3.3 Constraint with the Tool

The rupture resistance condition of a milling cutter constraint is written as following:

$$g_3 = \frac{8 k_s a_p f_z z V_c}{\Pi^2 D^3} \leq \tau_{max} \tag{17}$$

4 Mathematical Formulation

In this paper, the objective is to find the optimum cutting parameters in a single pass of face milling operation (rotational speed Ω, feed per tooth f_z and axial depth of cut a_p) to minimize the cutting time f_1, the cutting energy f_2, the surface roughness f_3 and the machining cost f_4 at the same time. In order to normalize the total objective function, an optimization of each function is elaborated to obtain f_1^* the minimum cutting time, f_2^* the minimum cutting energy, f_3^* the minimum surface roughness and f_4^* the minimum cutting cost. Our optimization problem is described as following:

$$\begin{cases} min(F) = \frac{f_1}{f_1^*} + \frac{f_2}{f_2^*} + \frac{f_3}{f_3^*} + \frac{f_4}{f_4^*} \\ s.c: \begin{cases} g_1 \leq f_{max} \\ g_2 \leq P_{max} \\ g_3 \leq \tau_{max} \end{cases} \end{cases} \tag{18}$$

The limit of the machine tool must be also considered as following:

$$\begin{cases} \Omega_{min} \leq \Omega \leq \Omega_{max} \\ f_{zmin} \leq f_z \leq f_{zmax} \\ a_{pmin} \leq a_p \leq a_{pmax} \end{cases} \tag{19}$$

5 Results and Discussions

To resolve the optimization problem, particle swarm algorithm (PSO) is used firstly to find f_1^*, f_2^*, f_3^* and f_4^* and secondly to find the minimum global objective function F. Indeed, PSO can solve a variety of difficult optimization problems and it is characterized with a few parameters to adjust, which makes it particularly easy to implement. Furthermore, research show that PSO algorithm has a better performance compared with other algorithms. In our study, each resolution is repeated 10 times to decrease the random effect of PSO algorithm. The tool and the workpiece materials are respectively carbide and steel. The parameters used during the simulation are summarized in Table 1.

The mono objective optimizations performed for the same milling process of only

Table 1 Simulation parameters

Parameters	Value
Workpiece length (mm)	100
Tool diameter (mm)	40
Radial depth of cut (mm)	20
Axial depth of cut range of variation $[a_p^{min}\ a_p^{max}]$ (mm)	[1, 4]
Feed per tooth range of variation $[f_z^{min}\ f_z^{max}]$ (mm/tooth)	[0.1; 0.6]
Rotational speed range of variation $[\Omega^{min}\ \Omega^{max}]$ (rpm)	[397,8; 2387]
Roughness parameters	$k = 1.001$, $x_1 = 0.0088$, $x_2 = 0.3232$, $x_3 = 0.3144$
Machine cost ($/min)	$k_0 = 0.3$
Tool cost ($)	$k_t = 6.87$
Energy cost ($/KWh)	$k_e = 0.13$
Specific pressure (N/mm^2)	$K_s = 2000$

one objective function results are recapitulate in Table 2. For each optimization we calculate the value of the others function based on the optimum cutting conditions.

The cutting parameters obtained by minimizing the cutting time are different from

Table 2 Optimization results of only one objective function

Model 1			Model 2			Model 3			Model 4		
$f_1^* = 1.075$ (s)			$f_2^* = 3.19 \times 10^2$ (J)			$f_3^* = 0.6$ (mm)			$f_4^* = 6.92$ ($)		
f_2 (J)	f_3 (mm)	f_4 ($)	f_1 (s)	f_3 (mm)	f_4 ($)	f_1 (s)	f_2 (J)	f_4 ($)	f_1 (s)	f_2 (J)	f_3 (mm)
1.063×10^3	1.3	22.96	6.43	1.63	6.92	10.2	4.46×10^3	9.64	6.28	3.19×10^2	1.63

ones obtained by minimizing surface roughness, cutting energy and cutting cost. Similar results are obtained for the cutting energy, the surface roughness and cutting cost. For this reason a global optimization of these four functions is elaborated as described in Eq. (18) in the next step. The results of the best solution obtained from the 10 resolution performed are summarized in Table 3.

Table 3 Optimization results of the global objective function

Minimization model 5	f_1 (s)	f_2 (J)	f_3 (mm)	f_4 ($)	*Min*(*F*)
Best solution	3.055	5.11 10^2	1.3	11.06	6.22

We conclude that the proposed optimization model (model 5 given by Eq. 18) ensure a balance between the minimum machining time, minimum machining energy, minimum machining cost and minimum of surface roughness. Indeed, compared to model 1, model 5 increases the machining time by 64.81% but decreases the cutting energy by 52%, the surface roughness are similar and the cutting cost is decreased by 52%. In comparison to model 2, it decreases the cutting time by 53%, the surface roughness by 20.24% but increases both the cutting energy and the cutting cost by 37%. When model 5 is compared to model 3, it decreases the cutting time by 70.04%, the cutting energy by 88% but increases the surface roughness by 54% and the cutting cost by 12%. Finally, compared to model 4, model 5 decreases the cutting time by 51.35% and the surface roughness by 18.75% but it increases the cutting energy and the cutting cost both by 37%. Those results prove that the proposed model 5 has a great efficiency to find a trade-off between the four objective functions in order to minimizing them.

6 Conclusion

In this paper, a mono objective optimization of a global model for minimizing cutting time, cutting cost, cutting energy and surface roughness is proposed and solved through PSO algorithm. A case study of single pass of face milling operation is conducted and search for the trade off solutions of minimizing cutting time, cutting cost, cutting energy and surface roughness. Three decision variables are taken into account such as rotational speed, axial depth of cut and feed per tooth. This work ameliorates the background described above by considering the surface roughness as an objective function and by considering the dynamic behavior of the cutting force during cutting energy modeling. As perspective, we propose to validate the obtained results from PSO algorithm by other results obtained from another algorithm such as Genetic Algorithm. We can also optimize a multi-pass face milling operation.

References

1. Albertelli P, Keshari A, Matta A (2016) Energy oriented multi cutting parameter optimization in face milling. J Clean Prod 137:1602–1618
2. Budak E (2006) Analytical models for high performance milling. Part I: Cutting forces, structural deformations and tolerance integrity. Int J Mach Tools Manuf 46:1478–1488
3. Dahmus J, Gutowski T (2004) An environmental analysis of machining. In: Proceedings of the 2004 ASME international mechanical engineering congress and RD&D exposition, Anaheim, California, USA

4. Hentati T, Barkallah M, Bouaziz S, Haddar M (2016) 1982. Dynamic modeling of spindle-rolling bearings systems in peripheral milling operations. J Vibroeng 18
5. Jang DY, Jung J, Seok J (2016) Modeling and parameter optimization for cutting energy reduction in MQL milling process. Int J Precis Eng Manuf Green Technol 3(1):5–12
6. Jin M, Tang R, Huisingh D (2015) Call for papers for a special volume on advanced manufacturing for sustainability and low fossil carbon emissions. J Clean Prod 87:7–10
7. Li C, Xiao Q, Tang Y, Li L (2016) A method integrating Taguchi, RSM and MOPSO to CNC machining parameters optimization for energy saving. J Clean Prod 135:263–275
8. Li C et al (2017) Selection of optimum parameters in multi-pass face milling for maximum energy efficiency and minimum production cost. J Clean Prod 140:1805–1818
9. Luan X, Zhang S, Cai G (2016) Optimal cutting parameters to reduce power consumption in face milling of a cast iron alloy for environmental sustainability. In: Sustainable design and manufacturing. Springer, Cham, pp 135–148
10. Mouzon G, Yildirim MB, Twomey J (2007) Operational methods for minimization of energy consumption of manufacturing equipment. Int J Prod Res 45:4247–4271
11. Tapoglou N, Mehnen J, Butans J, Morar NI (2016) Online on-board optimization of cutting parameter for energy efficient CNC milling. Procedia CIRP 40:384–389
12. Wang Z, Yuan J, Yin Z, Li C (2016) Study on high-speed cutting parameters optimization of AlMn1Cu based on neural network and genetic algorithm. Adv Mech Eng 8 (4):1687814016644126
13. Zarei O, Fesangharyb M, Farshia B, Jalili SR, Razfarb MR (2009) Optimization of multi-pass face-milling via harmony search algorithm. J Mater Process Technol 209:2386–2392

Investigation on the Effects of Recycling and Injection Parameters on Gloss Properties of Smooth Polypropylene Parts

Zaineb Baccouch[1(✉)], Souad Mbarek[2], Didier Perrin[3], Olivier Eterradossi[3], Bernard Monasse[4], Helene Garay[3], and Jean-Christophe Quantin[3]

[1] Laboratoire des Systems Electromécaniques, Ecole Nationale d'Ingénieurs de Sfax, Sfax, Tunisia
baccouch.zaineb@gmail.com
[2] Laboratoire de Mécanique de Sousse (LMS), LR11ES36, Ecole Nationale d'Ingénieurs de Sousse, Sousse, Tunisia
[3] Centre des Matériaux des Mines d'Alès (C2MA), Ecole des Mines d'Alès, Alès, France
[4] Centre de Mise en Forme des Matériaux (CEMEF), Ecole des mines de Paris, Paris, France

Abstract. The aim of this paper is to investigate the optical properties of the recycled polymer during numerous internal reprocess using experimental design. The process conditions (material temperature, mold temperature, injection rate) and recycling on the gloss and colorimetric properties of polypropylene containing 2 wt% of pigment was studied. Several injection parameters and cycles numbers must be tested. One most limit for this kind of study is the large number of experiments that requires longtime and significant investments. The idea is to vary three injection parameters (Temperature of material, Temperature of the mold, injection rate) for five injection cycles using statistical approach. The three variables (materiel temperature, mold temperature and injection flow) were investigated at three industrial used levels. The number of recycling varies from cycle 0 to cycle 4 at five levels. The complete matrix for screening was designed using D-optimal quadratic design. The experimental design was generated with the statistical software MODDE 10.1-Umetrics. A set of 45 experiments was carried out to determine the influence of injection parameters and recycling on the appearance properties of samples. The statistical software package Nemrodw® version 2007, LPRAI (Marseille, France) was used to analyze the experimental design.

Keywords: Recycling · Gloss properties · PP · Experimental design

1 Introduction

Appearance properties of objects are one of the most challenges of nowadays industries, to keep a place in a diversified market full of competition. To this commitment of quality was also added the necessity of recycling, especially for the plastics industry. In

F. Chaari et al. (eds.), *Advances in Materials, Mechanics and Manufacturing*,
Lecture Notes in Mechanical Engineering,
https://doi.org/10.1007/978-3-030-24247-3_25

a commercial production process, it is important to get rid of waste product. These are usually recycled to improve cost performance. However, this recycling process often brings unfavorable effects, e.g., an increase in appearance properties. To achieve this compromise between recycling and good esthetic of produces, studies on the effect of recycling on the appearance properties of polymers are needed. The effect of recycling on optical properties of polymer films has been few analyzed in the literature [1, 2] but no study has been done on the effect of recycling on the appearance properties of injected parts. Among other plastics, polypropylene (PP) is a commodity polymer product and used in large quantities for many applications. The main reasons for the success of PP are its quite good price/performance ratio, its excellent mechanical properties, and suitable optical characteristics [3]. The huge consumption of this polymer makes its recycling strategically very important for the environmental policy of industry [4]. On the other hand, appearance properties of polymer products are mostly determined by the processing parameters. A low cavity surface temperature makes the polymer melt freeze prematurely and consequently a frozen layer will be formed during filling process at the interface between the hot polymer melt and the cold mould cavity, which leads to a series of defects of the final moulded parts, such as flow mark [5, 6] weld mark [7, 8], swirl mark [9], roughness [10], low gloss [11], and low replication accuracy [12]. Zhang et al [13] showed that surface quality of plastic parts can be improved significantly by increasing mold cavity temperature. The gloss of both ABS/PMMA and ABS/PMMA/nano-$CaCO_3$ gradually increase with the increase of mold cavity temperature. The process conditions had a strong influence on the gloss development and on the color of pigmented PP. A better replication of the mold texture can be achieved at a lower melt viscosity at higher shear rates and higher mold temperatures. This gives a higher gloss in smooth regions. The gloss had a significant effect on the color: all the factors that contributed to an increase in gloss showed concomitant effects of increasing the color coordinate b* and of decreasing the lightness L*. Studies were published showing that the processing parameters may affect the gloss of injection molded parts [14], the mold temperature commonly being considered the more important parameter to be controlled. It was clearly shown that the mold temperature is the more relevant parameter and that a clear interaction exists between that parameter and the holding pressure.

From a physical point of view, the two quantitative descriptors of appearance (gloss and color) are the consequences of complex psychophysical phenomena of visual perception related to a situation in which the light reflected from the surface of an opaque sample is either predominantly in the specular direction (gloss) or diffuse in all directions (color) [15]. The four variables that primarily affect the gloss are the surface topography (or texture) [16], the wavelength and angle of the incident light and the refractive index of the material [17], whereas the color depends on the illumination conditions, the observation angle, the optical characteristics of the material, the amount of the colorant present, the surface topography [18], and the gloss [19]. Work on polymeric surfaces has shown a decrease in gloss with increased roughness [20].

In this paper, we mainly focus on the effect of process conditions (material temperature, mold temperature, injection rate) and recycling on the gloss and colorimetric properties of polypropylene containing 2 wt% of pigment using statistical approach.

2 Materials and Techniques

2.1 Materials

The PP homopolymer SABIC® PP 575P produced by SABIC Europe was used in the experiments. SABIC® PP 575P is an homopolymer for injection molding. A masterbatch supplied by Clariant was used to obtain a gray-beige complexion. The masterbatch was characterized with EDS analysis by scanning electron microscopy. The analyze show the presence of: $CaCO_3$ (calcium carbonate), TiO_2 (titanium dioxide) and Sb_2O_3 (Antimony trioxide). Blending of PP with 2 wt% of pigments (masterbatch) was performed in the PEP "plastics technical center" with a co-rotating intermeshing twin-screw extruder (Clextral EVOLUM HT 32, diameter 32 mm, centerline distance 21 mm) with a length to diameter ratio L/D 44. The screw profile is made of conveying and kneading elements, also using opposite pitch to ensure melting, mixing, shearing, and a good dispersion of the components. The extrusion process was carried out with a screw rotation speed of 350 rpm, at a temperature of 210 °C and a throughput of 30 kg h^{-1}. The colored pellets were also injection-molded to obtain samples for study.

2.2 Techniques

Material processing: The mold was supplied by CFO company specialized on "Manufacturing Design Tools". The mold design was realized in C2MA to obtain samples of 100 mm × 100 mm with a thickness of 2 mm. the samples show two different faces, a polished mirror surface and a rough surface. All specimens were prepared on a Krauss Maffei KM50-180CX injection molding machine. The maximum clamping force is 50 tons. During experiment, corresponding to each set of experimental parameters, the 5 first injected samples are thrown away to ensure that the process was stable. PP was injected from zero to five steps of recycling. Samples are grinding with a Cutting Mill SM 300 to be reinjected.
Colorimetry: In a uniform color scale, the differences between points plotted in the color space correspond to visual differences between the colors plotted. The L* axis is the light-dark axis, and thus gives a measure of the relative brightness of the sample ranging from total black (L* = 0) to the white (L* = 100). The a* axis is the red-green axis (positive a* = red, negative a* = green), and the b* axis is the blue-yellow axis (positive b* = yellow, negative b* = blue) [21] (Fig. 1).
Brightness: In terms of physical measurements, especially in the industrial field, gloss of materials is considered only at specular angles using a glossmeter [22]. Glossmeters are built to afford three specular geometries (Fig. 2). The geometry is chosen based on the gloss level observed, selecting the 20 specular reflection for high-gloss materials and 85 for quasi-matt samples, while 60 is used for samples with an intermediate, satin-like appearance.
Statistical approach: In a purpose of understanding the effect of injection parameters and recycling on the appearance properties of polymers, several injection parameters and cycles numbers must be tested. One main limit for this kind of study is the large number of experiments that requires longtime and significant investments. The idea is to vary three injection parameters temperature of material

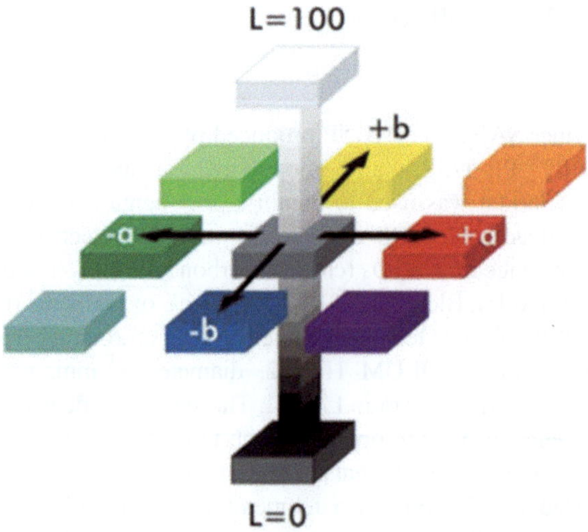

Fig. 1 (L*, a*, b*) representing color space

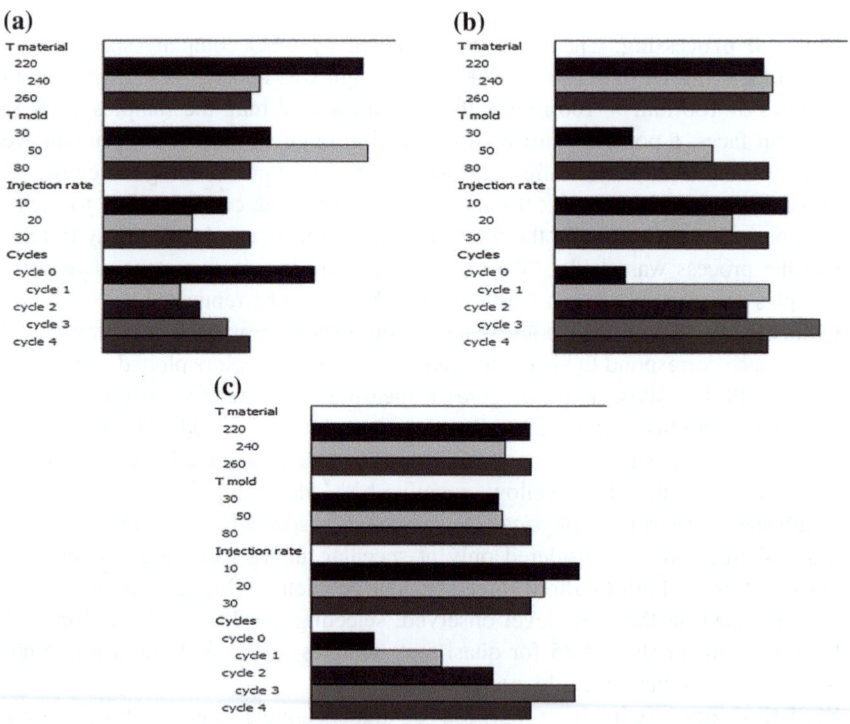

Fig. 2 Graphic representation of the difference in weight of the factors on total effect for the gloss response **a** R20, **b** R60, **c** R85

(Tmaterial), Temperature of the mold (Tmold) and the injection rate for five injection cycles. The work material selected for the study was polypropylene because of its large use and the need to recycle it with the maintaining of good aspect properties.

The effects of four factors known to influence injection conditions, material temperature, mold temperature, injection flow and number of injection cycles were studied using a statistical approach. The three variables (materiel temperature, mold temperature and injection flow) were investigated at three industrial used levels (materiel temperature: 220, 240, 260 °C, mold temperature: 30, 50, 80 °C and injection flow: 10, 20, 40 cm^3/s). The number of recycling varies from cycle 0 to cycle 4 at five levels. The complete matrix for screening was designed using D-optimal quadratic design. The experimental design was generated with the statistical software MODDE 10.1-Umetrics. A set of 45 experiments was carried out to determine the influence of injection parameters and recycling on the appearance properties of samples. The statistical software package Nemrodw® version 2007, LPRAI (Marseille, France) was used to analyze the experimental design.

The response approach involving a D-optimal quadratic design was adopted for studied the effect of injection parameters and recycling on the final response (colorimetric and gloss). In fact, A D-optimal quadratic design is a computer aided design which contains the best subset of all possible experiments. Depending on a selected criterion and a given number of design runs, the best design is created by a selection process.

Each variable (material temperature, mold temperature, injection rate) was studied at three different levels (1, 2, and 3). The four recycling cycles was studied with four levels (1, 2, 3, and 4). All variables were taken at a central coded value of zero. The set of levels of the variables is summarized in Table 1.

Table 1 D-optimal plan variables used in the area of interest (Design of Experiments—DOE)

Type de variable	Name(Pi)	Value		
Discrete block	Number of cycle	0		
		1		
		2		
		3		
		4		
Continue	T material (°C)	220	240	260
	Tmold (°C)	30	50	80
	Injection rate (cm^3/s)	10	20	40

Each variable was coded between −1 and 1 according to Eq. (1):

$$X_i = 2\frac{P_i - \frac{P_{imax}+P_{imin}}{2}}{P_{imax} - P_{imin}} = 2\frac{P_i - \overline{P_i}}{\Delta P_i} \qquad (1)$$

with P_i, the variable to be coded between P_{imin} and P_{imax} respectively corresponding to the minimum and maximum of the experimental variables. The use of reduced centered variables has the advantage of being able to generalize the theory of the experiments plans whatever the factors or the fields of studies retained.

The responses have been adapted in order to know the final evolution of the colorimetric and brightness parameters according to the quadratic D-optimal model described above and corresponding to Eq. (2).

The minimum and maximum range of variables investigated and the full experimental plan with respect to their actual and coded forms. The analysis of the data was carried out to obtain an empirical model defining the response Y (Eq 1), which is the value of colorimetric (L*, a*, b*) or gloss (R20, R60, R85).

$$\begin{aligned} Y = b_0 + b_{1A}.X_{1A} + b_{1B}.X_{1B} + b_{2A}.X_{2A} + b_{2B}.X_{2B} + b_{3A}.X_{3A} \\ + b_{3B}.X_{3B} + b_{4A}.X_{4A} + b_{4B}.X_{4B} + b_{4C}.X_{4C} + b_{4D}.X_{4D} \end{aligned} \qquad (2)$$

Yi was the responses in terms of colorimetry and brightness. This result was of three separate tests and reflect the colorimetry values L*, a*, b* and brightness respectively at angles R20, R60, and R85.

3 Results and Discussion

Results in terms of gloss and colorimetric responses are analyzed with the statistical software package Nemrodw-version 2007, LPRAI (Marseille, France) and shown respectively in Figs. 2 and 3. The reference is taken on the value of the highest level (black bar). The influence of the injection temperature (220, 240 and 260 °C), the mold temperature (30, 50 and 80 °C), the injection rate (10, 20 and 40 cm^3/s) and the number of injection cycle (cycle 0 to cycle 5) on the gloss of different PP samples with smooth surface is shown in Fig. 2.

Figure 3 shows the effect of varying injection parameters and number of cycles on gloss (R20, R60, and R85). It can be observed that the recycling number change the gloss properties by varying R20, R60 and R85. These variations are not in the same way because of recycling lead to a decrease in R20 and an increase in R60 and R85. The effect of material temperature is observed only for R20 (Fig. 2a) for which the increase in material temperature lead to a decrease in R20. The modification of the mold temperature leads to variations in R20 and R60. The increase in mold temperature leads to an increase in R20 and R60. The interaction between the effect of material temperature and mold temperature lead to the increase of R20 at 80 °C of mold temperature. The injection rate does not significantly affect the gloss properties. In general, the number of injection cycles and the mold temperature has the greatest effect on the gloss level. For colorimetric response (Fig. 3) we have the same important effect

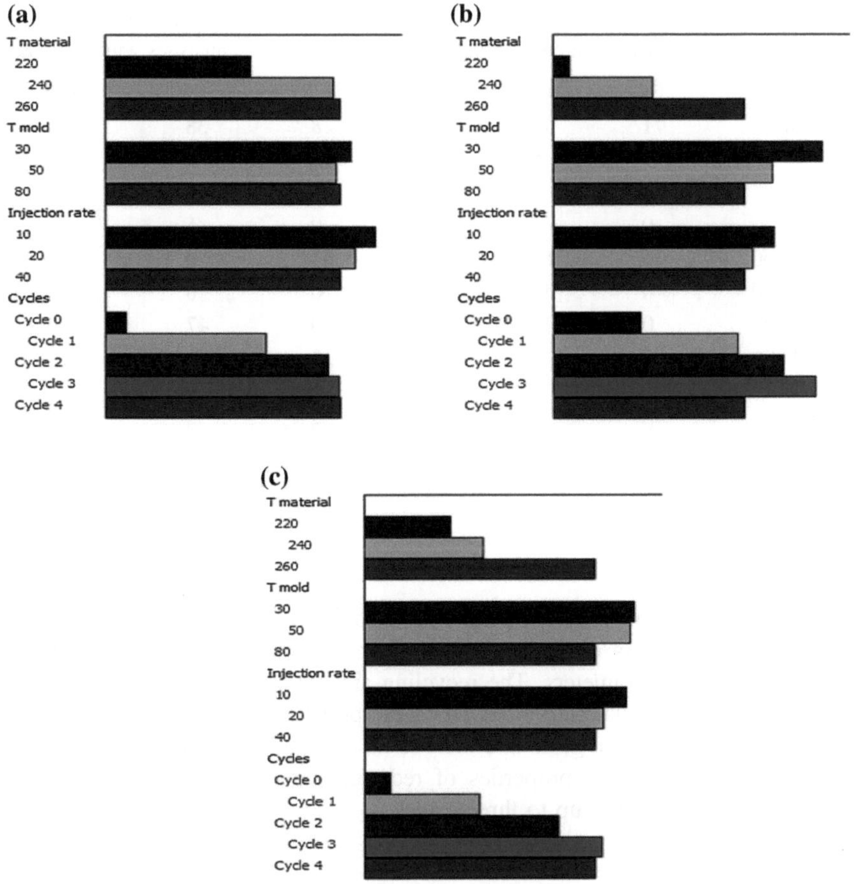

Fig. 3 Graphic representation of the difference in weight of the factors on total effect for the colorimetric response **a** L*, **b** a*, **c** b*

of the recycling and the material temperature on (L*, a*, b*) that increases with increasing the number of cycles and the material temperature. The effect of mold temperature and injection rate is opposite and leads to a decrease in (L*, a*, b*).

We can note that colorimetric and a gloss property depends on changes in injection conditions and number of recycling cycles. The recycling and material temperature have the most important effect on colorimetric. The variations of gloss are obtained by changing the number of cycles, the material temperature and the mold temperature. Table 2 illustrate the pareto effect of all parameters. So, we notice that the numbers in bold highlighted the most important effect on gloss measurements. The studies of the appearance properties of the injection molded samples revealed that changes in injection parameters and number of cycles had a significant influence on colorimetric and gloss. Generally, appearance properties are related to the surface roughness. The differences in optical and appearance properties between samples were presumed to arise from the variation in the surface topography in according to the literature.

Table 2 The effect of injection parameters and number of cycles on the colorimetric and specular gloss measurements (Pareto individual effect from response analyses with Nemrodw)

	L*	a*	b*	R20	R60	R85
Cycle 0/4	**71**	14	**40**	8	**38**	**58**
Cycle ¼	9	0	14	11	0	20
Cycle 2/4	0	2	1	5	0	3
Cycle 3/4	0	7	0	0	5	5
T_{mat} 220/260	16	**50**	26	**32**	0	0
T_{mat} 240/260	0	13	15	0	0	2
T_{mold} 30/80	0	11	2	0	**47**	3
T_{mold} 50/80	0	1	0	**32**	7	2
Flow 10/40	2	1	0	1	0	7
Flow 20/40	0	0	0	8	3	0

4 Conclusion

The optical properties of recycled parts can be maintained during numerous internal recycling. The variations of properties are in the same range that is observed by changing processing parameters. The recycling and the material temperature are the most important factors changing the colorimetric properties of the material. The mold temperature affects only the gloss and there is no change in properties by varying the injecting flow. Also, aspect properties of recycled parts can be maintained during numerous internal recycling up to three cycles.

References

1. Allen NS, Hardy SJ, Jacobine AF, Glaser DM, Yang B, Wolf D, Catalina F, Navaratnam S, Parsons BJ (1991) Photochemistry and photopolymerization activity of perester derivatives of benzophenone. J Appl Poly Sci 42:1451
2. Baccouch Z, Mbarek S, Jaziri M (2017) Experimental investigation of the effects of a compatibilizing agent on the properties of a recycled poly(ethylene terephthalate)/polypropylene blend. Polym Bull 74:839
3. Moore EP (1996) Polypropylene handbook: polymerization, characterization, properties, processing, applications. Hanser, Munich, Germany
4. Vilaplana F, Karlsson S (2008) Quality concepts for the improved use of recycled polymeric materials: a review. Macromol Mater Eng 293:274
5. Tredoux L, Satoh I, Kurosaki Y (1999) Investigation of wave-like flow marks in injection molding: flow visualization and micro-geometry. Polym Eng Sci 39:2233–2241
6. Yoshii M, Kuramoto H, Kawana T (1996) The observation and origin of micro flow marks in the precision injection molding of polycarbonate. Polym Eng Sci 36:819–826
7. Fellahi S, Meddad A, Fisa B, Favis BD (1995) Weldlines in injection-molded parts: a review. Adv Polym Technol 14:169–195

8. Xie L, Ziegmann G (2008) A visual mold with variotherm system for weld line study in micro injection molding. Microsyst Technol Micro-Nanosyst-Inform Storage Process Syst 14:809–814
9. Lee J, Turng LS (2010) Improving surface quality of microcellular injection molded parts through mold surface temperature manipulation with thin film insulation. Polym Eng Sci 50:1281–1289
10. Liu SJ, Chang JH (2000) The occurrence of surface roughness in gas assist injection molded nylon composites. Polym Composite 21:322–331
11. Oliveira MJ, Brito AM, Costa MC, Costa MF (2006) "Gloss and surface topography of abs: a study on the influence of the injection molding parameters. Polym Eng Sci 46:1394–1401
12. Theilade UA, Hansen HN (2007) Surface microstructure replication in injection molding. Int J Adv Manuf Technol 33:157–166
13. Zhang A, Zhao G Guan Y (2015) Effects of mold cavity temperature on surface quality and mechanical properties of nanoparticle-filled polymer in rapid heat cycle molding. J Appl Polym Sci
14. Dawkins E, Engelmann P, Horton K, Monfore M (1998) Color and gloss-the connection to process conditions J Inject Mold Technol 1:1
15. Hunter RS, Harold RW (1987) The Measurement of appearance. Wiley, New York
16. Withehouse DJ, Bowen DK, Venkatesh VC, Lonardo P, Brown CA (1994) Cirp annals 43:541
17. Donald B, Mathew R (1988) SPE Antec Tech Paper 34:18
18. Huff K (1994) Visual assessment and practical colorimetry in the plastic industry. Bayer AG, Leverkusen, Germany
19. Dalal EN, Natale-Hoffman KM (1999) The effect of gloss on color. Color Res Appl 24:369
20. Wang L, Huang T, Kamal MR, Rey AD, Teh J (2000) Surface topography and gloss of polyolefin blown films. Polym Eng Sci 40:747–760
21. Lange Dr (1981) Color difference measuring instrument. Micro color, operating instructions, 4 edn. BDA 163
22. ISO 2813 (1994) Paints and varnishes. Measurement of specular gloss of nonmetallic paint films at 20_, 60_ and 85_

MIX
Papier aus verantwortungsvollen Quellen
Paper from responsible sources
FSC® C105338

If you have any concerns about our products,
you can contact us on
ProductSafety@springernature.com

In case Publisher is established outside the EU,
the EU authorized representative is:
Springer Nature Customer Service Center GmbH
Europaplatz 3, 69115 Heidelberg, Germany

Printed by Libri Plureos GmbH
in Hamburg, Germany